Praise for
YOU ARE THE UNIVERSE

"I am often asked if Deepak Chopra really believes the many controversial and provocative ideas he espouses in his many writings. Now that I have gotten to know him, I can answer unequivocally in the affirmative, and there is no better encapsulation of his scientific worldview than *You Are the Universe,* which he coauthored with the highly respected physicist Menas Kafatos, my colleague at Chapman University. If you want to understand the worldview in which human consciousness is primary, and how that perspective can be defended through science, this is the book to read. In my own journey to better understand Deepak and his worldview, this book was the most enlightening path I took."

—MICHAEL SHERMER, PH.D., publisher of *Skeptic* magazine; monthly columnist of *Scientific American;* Presidential Fellow at Chapman University; author of *The Moral Arc, The Believing Brain,* and *Why People Believe Weird Things*

"As a teenager, I used to find it rather curious that people regard their thoughts and emotions as integral to who they are, but their perceptions as something totally beyond themselves. The world we perceive is, after all, part of our mental life just like our thoughts and emotions. In this book, Deepak and Menas take this seemingly innocent idea to cosmic heights, revealing its true force and significance. They do it intelligently, in a scientifically well-informed manner, and with good taste. The result is delightful."

—BERNARDO KASTRUP, PH.D., author of *Why Materialism Is Baloney, Brief Peeks Beyond,* and *More Than Allegory*

"*You Are the Universe* could have been spelled *Youniverse,* for not only are 'you' in the universe, 'you' are at the start of it all. Chopra and Kafatos have put together a well-written and, as far as any scientist today knows, a completely accurate exploration of how the mystery of subjective consciousness provides the basis for material reality as it is presently understood. I highly recommend this for those who are curiously alive."

—FRED ALAN WOLF, PH.D., aka Dr. Quantum®, theoretical physicist; author of *The Spiritual Universe*, National Book Award–winning *Taking the Quantum Leap*, *Time Loops and Space Twists*, and many other books

"The latest masterpiece by Deepak is a joint oeuvre with cosmologist Menas Kafatos. It addresses all the most important questions we can ask of ourselves and of science. Questions like who are we, and why are we here—with the science to back our answers. This is the 'new paradigm' we have been talking about!"

—ERVIN LASZLO, author of *What Is Reality: The New Map of Cosmos, Consciousness, and Existence*

"In this interesting book, an astrophysicist is uniquely teaming up with a medical doctor. They present a novel, and I dare to say, revolutionary 'paradigm' that has to make us all reconsider our ideas about our place in the universe. It will shake stagnated waters in the shortsighted beliefs of many. It will also make us think and wonder about our real relationship with the cosmos."

—KANARIS TSINGANOS, director and president of the governing board of the National Observatory of Athens; professor of astrophysics, astronomy, and mechanics in the department of physics, University of Athens (Greece)

"*You Are the Universe* discusses the most important aspect of consciousness studies, that is, 'Does the mind create reality?' This book raises a lot of such fascinating issues that may create an environment of new debate."

—Sisir Roy, T.V. Raman Pai Chair at the National Institute of Advanced Studies, IISc campus, Bangalore; (former) professor of physics and applied mathematics unit at the Indian Statistical Institute, Kolkata, India

"*You Are the Universe* brings the usual gracious clarity of all of Deepak Chopra's writings together with the insights of physicist Menas Kafatos to elucidate the most profound and pressing questions at the frontiers of contemporary science. Weaving Dr. Chopra's expertise regarding biological systems with Professor Kafatos's work in quantum physics, geophysics, and cosmology, they illuminate the realms where all the most successful contemporary sciences come to the edge of what can be explained with the vital lights from their own lifetimes of deep spiritual practice. The result is no clash of competing perspectives, but a rich, synergistic tapestry of great wisdom, beauty, and comfort for our culture. As such, *You Are the Universe* is their great and generous gift to each of us."

—Neil Theise, M.D., professor of pathology at the Icahn School of Medicine at Mount Sinai

You Are the

DISCOVERING YOUR COSMIC SELF

Universe

AND WHY IT MATTERS

DEEPAK CHOPRA, M.D., and

MENAS C. KAFATOS, Ph.D.

HARMONY

BOOKS • NEW YORK

Published in the United States by Harmony Books,
an imprint of the Crown Publishing Group, a division
of Penguin Random House LLC, New York.
crownpublishing.com

Harmony Books is a registered trademark, and the Circle
colophon is a trademark of Penguin Random House LLC.

Library of Congress Cataloging-in-Publication Data
is available upon request.

ISBN 978-0-307-88916-4
Ebook ISBN 978-0-307-95185-4
International Edition ISBN 978-0-8041-8992-7

Printed in the United States of America

Book design by Lauren Dong
Jacket design by Pete Garceau

10 9 8 7

First Edition

ACKNOWLEDGMENTS

For a fruitful collaboration, especially when a book is as complexly woven as this one, many thanks are deserved.

We are extremely grateful to a friend and outstanding physicist, Leonard Mlodinow of Caltech, who gave our manuscript a close critical examination. Similar thanks go to the knowing, talented science writer Amanda Gefter. They served to ensure that our science was as close to note perfect as possible, even when we ventured into controversial areas that challenge mainstream science.

The study of consciousness has moved from being a minor consideration in hard science to a major investigation. We have gained much from three major conferences on the subject and to their tireless organizers:

Stuart Hameroff, a leading pioneer in the field. He heads the invaluable Science and Consciousness Conference: http://consciousness.arizona.edu/.

Maurizo and Zaya Benazzo, who are the conference founders and organizers of SAND, a science and nonduality conference of international reach and importance: https://www.scienceandnonduality.com/.

Sages and Scientists Symposium, hosted by the Chopra Foundation: www.choprafoundation.org.

On the publishing arm, we were extremely fortunate in the team who made this book possible, beginning with our patient, dedicated editor, Gary Jansen, as well as the working team at Harmony: Aaron Wehner, publisher; Diana Baroni, vice president and editorial director; Tammy Blake, vice president and director of publicity; Julie Cepler, director of marketing; Lauren Cook, senior publicist; Christina Foxley, senior marketing manager; Jenny Carrow and Christopher Brand, our jacket design team; Elizabeth Rendfleisch, director of interior design; Heather Williamson, production manager; and Patricia Shaw, senior production editor. Thanks also go to the executives, Maya Mavjee, president and publisher of the Crown Publishing Group, and again, Aaron Wehner, senior vice president and publisher of Harmony Books.

The coauthors also have important individuals to thank.

FROM MENAS:

My family has played a fundamental role in shaping who I am as a person and a scientist, starting with my parents, Constantine and Helen, who taught me to respect others and follow good principles in life; my oldest brother, Anthony, who always stood by me and protected me; and my brother Fotis, whom I followed at Cornell and who showed me the first steps of being a scientist. My uncle George Xiroudakis instilled in me a love for mathematics. My thesis advisor at MIT, Philip Morrison, gave me basic understandings and enthusiasm for astrophysics and cosmology. I offer gratitude to all the great professors at MIT, Cornell, and Harvard with whom I studied.

My deepest thanks also to my wife, Susan Yang. You have

always been supportive and stood by me as my horizons expanded. My three sons, Lefteris, Stefanos, and Alexios, give me such meaning and deep regard as a father. I gratefully acknowledge my great friends and extended family in the U.S., South Korea, and Greece, who believe in the same dreams, whatever our differences. I see you all as part of me. Finally, my science and philosophy would have been nothing without Niels Bohr, all the great quantum physicists, and my spiritual teacher.

FROM DEEPAK:

For everything you give with unstinting love, I offer thanks to my wife, Rita, our children, Gotham and Mallika, and our grandchildren, who offer enormous optimism about the future.

Both authors want to thank the great team at the Chopra Center, most especially Carolyn, Felicia, and Gabriela Rangel, the family in a family that handles the drama and details without which this book would have been impossible.

CONTENTS

PREFACE

YOU AND THE UNIVERSE ARE ONE

There's one relationship in your life—in everyone's life—that has been kept a secret. You don't know when it began, and yet you depend upon it for everything. If this relationship ever ended, the world would disappear in a puff of smoke. This is your relationship to reality.

A huge number of things must come together perfectly to construct reality, and yet they do so completely out of sight. Think of sunlight. Obviously, the sun can't shine unless stars exist, since our sun is a medium-size star floating beyond the center of the Milky Way, our home galaxy. There are few secrets left to discover about how stars form, what they are made of, and how light is produced in the incredibly hot cauldron at the core of a star. The secret lies elsewhere. As sunlight travels 93 million miles to Earth, it penetrates the atmosphere and lands somewhere on the planet. In this case, the only somewhere we're interested in is your eye. Photons, the packets of energy that carry light, stimulate the retina at the back of your eye, starting a chain of events that leads to your brain and the visual cortex.

The miracle of vision lies in the mechanics of how the brain processes sunlight, that much is clear. Yet the step that matters the most, converting sunlight into vision, is totally mysterious. No matter what you see in the world—an apple, cloud, mountain, or tree—sunlight bouncing off the object makes it visible. But how? No one really knows, but the secret formula contains sight, because seeing is one of the basic ways of knowing that an object is real.

What makes seeing totally mysterious can be summed up in a few undeniable facts:

Photons are invisible. They aren't bright, even though you see sunlight as bright.

The brain has zero light inside it, being a dark mass of oatmeal-textured cells enveloped in a fluid not terribly different from seawater.

Because there is no light in the brain, there are no pictures or images, either. When you imagine the face of a loved one, nowhere in the brain does that face exist like a photograph.

At present no one can explain how invisible photons being converted to chemical reactions and faint electrical impulses in the brain creates the three-dimensional reality we all take for granted. Brain scans pick up the electrical activity, which is why an fMRI contains patches of brightness and color. *Something* is going on in the brain. But the actual nature of sight is mysterious. One thing is known, however. The creation of sight is done by you. Without you, the entire world—and the vast universe extending in all directions—can't exist.

Sir John Eccles, a neurologist and Nobel laureate, declared, "I want you to realize that there exists no color in the natural world, and no sound—nothing of this kind; no textures, no patterns, no beauty, no scent." What Eccles means is that all the qualities of nature, from the luxurious scent of a rose to the sting of a wasp and the taste of honey, is produced by human beings. It's a remarkable statement, and nothing can be left out. The most distant star, billions of light-years away, has no reality without you, because everything that makes a star real—its heat, light, and mass, its position in space and the velocity that carries it away at enormous speed—requires a human observer with a human nervous system. If no one existed to experience heat, light, mass, and so on, nothing could be real as we know it.

That's why the secret relationship is the most important one you have or will ever have. You are the creator of reality, and yet you have no idea how you do it—the process is effortless. When you see, light gains its brightness. When you listen, air vibrations turn into audible sound. The activity of the world around you in all its richness depends upon how you relate to it.

This profound knowledge isn't new. In ancient India, the Vedic sages declared *Aham Brahmasmi*, which can be translated as "I am the universe" or "I am everything." They arrived at this knowledge by diving deep into their own awareness, where astonishing discoveries were made. Lost to memory are Einsteins of consciousness whose genius was comparable to the Einstein who revolutionized physics in the twentieth century.

Today we explore reality through science, and there cannot be two realities. If "I am the universe" is true, modern science must offer evidence to support it—and it does. Even though mainstream science is about external measurements, data, and experiments, which build a model of the physical world rather than the inner world, there are a host of mysteries that measurements, data, and experiments cannot fathom. At the far frontier of time and space, science must adopt new methods in order to answer some very basic questions, such as "What came before the big bang?" and "What is the universe made of?"

We present nine of these questions, the biggest and most baffling riddles that face science today. Our aim is not to put just another popular science book in readers' hands. We have an agenda, which is to show that this is a participatory universe that depends for its very existence on human beings. There is a growing body of cosmologists—the scientists who explain the origin and nature of the cosmos—developing theories of a completely new universe, one that is living, conscious, and evolving. Such a universe fits no existing standard model. It's not the cosmos of quantum physics or the Creation described as the work of an almighty God in the book of Genesis.

A conscious universe responds to how we think and feel. It gains its shape, color, sound, and texture from us. Therefore, we feel the best name for it is the *human universe*, and it is the real universe, the only one we have.

Even if you are new to science or have little interest in it, you can't help but be interested in how reality works. How you view your own life matters to you, of course, and everyone's life is embedded in the matrix of reality. What does it mean to be human? If we are insignificant specks in the vast black void of outer space, that reality must be accepted. If, instead, we are creators of reality living in a conscious universe that responds to our minds, that reality must be accepted. There is no middle ground and no second reality that can be chosen because we happen to like it better.

So let the journey begin. Every step of the way we will let you be the judge. For every question like "What came before the big bang?" you'll read about the best answers modern science can offer, followed by why these answers haven't been satisfactory. This opens the way for entirely new explorations into a universe where answers come from everyone's experience. This is probably the greatest surprise of all, that the control room for creating reality exists in the experiences everyone is having every day. Once we unfold how the creative process works, you will arrive at a completely different view of yourself than before. Science and spirituality, the two great worldviews in human history, both contribute to the ultimate goal, which is to discover what is "really" real.

A disturbing truth is dawning all around us. The present-day universe has not worked out the way it was supposed to. Too many unsolved riddles have piled up. Some are so baffling that even imagining how to answer them is in doubt. There's an opening for a totally new approach, what some call a paradigm shift.

A paradigm is the same as a worldview. If your paradigm or worldview is based on religious faith, a Creation needs a Creator, a divine agent who arranged the astonishing intricacies of the cosmos. If your paradigm is based on the values of the eighteenth-century Enlightenment, the Creator may still exist, but he has no business with the everyday workings of the cosmic machinery—he's more like a watchmaker who set the machine going and walked away. Paradigms keep shifting, driven by human curiosity and, for the past four hundred years, viewed through the lens held up by science. At the moment, the paradigm that dominates science posits an uncertain, random universe devoid of purpose and meaning. For anyone working within this worldview, progress is constantly being made. But we must remember that to a devout Christian scholar in the eleventh century, progress toward God's truth was constantly being made, too.

Paradigms are self-fulfilling, so the only way to cause radical change is to jump out of them. That's what we intend to do in this book, to jump from an old paradigm into a new one. But there's a hitch. New paradigms aren't simply pulled down from the shelf. They must be put to the test, and this is done by asking a simple question: Is the new worldview better at explaining the mystery of the universe than the old one? We believe that the human universe *must* prevail. It's not an add-on to any existing theory.

If the human universe exists, it must exist for you as an individual. The present-day universe is "out there," spanning immense distances and having little or no connection to how you live your daily life. But if everything you see around you needs your participation, then you are touched by the cosmos every minute of the day. To us, the biggest mystery is how human beings create their own reality—and then forget what they did. We offer our book as a guide to remembering who you really are.

The shift into a new paradigm is happening. The answers

offered in this book are not our invention or eccentric flights of fancy. All of us live in a participatory universe. Once you decide that you want to participate fully, with mind, body, and soul, the paradigm shift becomes personal. The reality you inhabit will be yours to either embrace or change.

No matter how many billions are spent on scientific research, no matter how fervently religionists keep faith with God, what finally matters is reality. The case for the human universe is very strong; it's part of the paradigm shift unfolding all around us. The reason we say "You are the universe" is that nothing less than that is the truth.

OVERVIEW

THE DAWN OF A HUMAN UNIVERSE

There is a photograph of Albert Einstein standing beside the most famous man in the world, who happened to be the great comedian Charlie Chaplin. In 1931, Einstein was touring Los Angeles, and a chance encounter at Universal Studios led to an invitation to attend the premiere of Chaplin's new movie *City Lights.* Both men are dressed in tuxedos and smiling broadly. It's astonishing to think that Einstein was the second-most-famous man in the world.

He didn't owe his worldwide fame to the fact that everyday people understood his theories of relativity.* Einstein's theories dwelt in a realm far above everyday life, and that in itself created awe. British philosopher and mathematician Bertrand Russell wasn't trained in physics; when Einstein's ideas were explained to him, he was astounded and burst out, "To think I have spent my life on absolute muck." (Russell went on to write a brilliant explanation for laymen, *The ABC of Relativity.*)

In some way relativity had toppled both time and space; the average person could grasp that much. $E = mc^2$ was the most famous equation in history, but what it meant didn't touch

* Although commonly referred to as the Theory of Relativity, Einstein issued his revolutionary idea in two stages, first as the Special Theory of Relativity in 1905, then as the grander General Theory of Relativity in 1915.

everyday life, either. People went about their daily existence as if none of Einstein's deep thinking mattered, not in practical terms.

But that assumption has turned out to be wrong.

When Einstein's theories toppled time and space, something real happened—the fabric of the universe was torn apart and then rewoven into a new reality. What few understood was that Einstein imagined this new reality; he didn't work with mathematics on a chalkboard. From childhood he possessed a remarkable ability to picture difficult problems in his head. As a student he'd try to visualize what it would be like to travel at the speed of light. The speed of light had been calculated at 186,000 miles per second, but Einstein felt that light contained something quite mysterious that hadn't been discovered. What he wanted to know was not the properties of light or what light was like as a physicist studied it, but what the *experience of riding a beam of light* would be like.

For example, the foundation of relativity is that all observers measure the same speed of light, even if they are moving at different speeds, away from each other or toward each other. This implies that nothing in the physical universe can travel faster than the speed of light, so imagine that you are traveling at essentially the speed of light and you throw a baseball in the direction you're traveling. Would it leave your hand? After all, your speed is already at the absolute limit; no extra speed can be added. If the baseball did leave your hand, how would it behave?

Once he got a mental picture of a problem, Einstein looked for an equally intuitive solution. What makes his solutions so fascinating—especially for our purposes—is how much imagination was being applied. For example, Einstein imagined a body in free fall. For someone having such an experience, there would appear to be no gravity. If he took an apple out of his pocket and let go of it, the apple would float in the air beside him, again making it seem that there was no gravity.

Once Einstein saw this in his mind's eye, he had a revolutionary thought: maybe there *is* no gravity in such a situation. Gravity had always been considered a force acting between two objects, but he saw it as nothing more than curved space-time, implying that space and time would be affected in the presence of mass. And that curved space-time, in the vicinity of collapsed objects such as black holes, would result in time stretching to a stop as seen by distant observers. Yet someone located at the falling object wouldn't see anything out of the ordinary. Demoting gravity as a force was one of relativity's most shocking features.

We can see Einstein's visualization in action when astronauts are trained in weightless conditions inside an airplane. The camera shows them floating in midair, completely free of gravity and, exactly as Einstein predicted, any unattached object inside the aircraft is also weightless. What the camera doesn't show is that to achieve zero gravity, the plane is accelerating rapidly in free fall, enough to counteract Earth's gravitational field. As relativity predicted, speed turns gravity into a changeable condition.

If gravity as a force is mutable, what about other things we take for granted as fixed and reliable? Einstein made another crucial breakthrough regarding time. In place of absolute time, which was taken as a given prior to relativity, he discovered that time is affected by an observer's frame of reference and also by being close to a strong gravitational field. This is known as time dilation. The clocks on the International Space Station appear to an astronaut to be running perfectly normally, while in relationship to clocks on Earth, they are slightly fast. A traveler nearing the speed of light wouldn't notice that the clocks on his spacecraft are acting any differently, but to an observer on Earth, they would appear to be slowing down. Clocks positioned close to a strong gravitational field run slower as viewed from far away.

Relativity shows us that there is no universal time. Clocks all over the universe cannot be synchronized. As an extreme example, a spaceship nearing a black hole would be affected by the black hole's immense gravitational pull, so much so that to an observer on Earth, the clocks on the spaceship would drastically slow down, actually taking an infinite time to cross the horizon of the black hole and be sucked inside. Meanwhile, for the crew falling into the black hole, time would run normally until in short order they would be crushed by its immense gravitational pull.

Although these effects have been known for a century, something new has occurred in our time—relativity actually matters in daily life. On Earth, clocks tick slower than in empty space far away from gravity. So, as clocks pull away from Earth's gravity, they speed up or, more correctly, they appear to, which means that the satellites used for GPS coordinates have faster clocks than the ones down here. When you ask the GPS device in your car to locate where you are, the answer would be off, if only by a little, unless the clocks on the GPS satellite were adjusted to match Earth time. ("A little" would be enough to mistake your location by several blocks, a disastrous error for a mapping and guidance system.)

Einstein's visual images began his journey to the Special Theory of Relativity, and for our purposes, that's critically important. He himself was amazed when his purely mental work turned out to match how nature really works. But everything the theory predicted, including black holes and the slowing of time in the presence of large gravitational forces, has come true. Einstein realized that time, space, matter, and energy were interchangeable. This single idea deposed the normal world of the five senses with its claim that nothing we see, hear, taste, touch, and smell is reliable.

You can do your own visualization to prove this fact to yourself. See yourself sitting on a train moving down the tracks. You look out the window and notice that a second train is traveling beside you on a second, parallel track. This second train isn't moving forward, however, so according to your eyes, it must be standing still. But your eyes are lying, since in reality your train and the second train are moving at the same speed relative to the platform. Mentally, we all adjust to the lies our senses tell us. We adjust to the lie that the sun rises in the east and sets in the west. As a fire truck whizzes past, its siren rises in pitch as it approaches and decreases in pitch as it races off into the distance. But mentally we know that the siren's sound hasn't changed. The rising and falling was a lie told by our ears.

Each sense is equally unreliable. If you tell someone that you are about to stick their hand into a bucket of scalding water, but instead you plunge the hand into ice water, most people will cry out as if the water was hot. A mental expectation causes the sense of touch to deliver a false picture of reality. So the relationship between what you think and what you see works two ways. Your mind can misinterpret what you see or your eyes can tell your mind a false story. (We're reminded of an incident that happened to an acquaintance. When he came home from work, his wife told him that there was a huge spider in the bathtub and begged him to get rid of it. He marched upstairs and pulled the shower curtain back. From downstairs his wife heard him shriek when he saw what he thought was the world's hugest spider. But in fact, it being April Fool's Day, she had put a live lobster in the bathtub!)

If the mind can fool the senses and the senses can fool the mind, reality becomes suddenly less substantial. How can we rely on an external "reality" if it's affected by how we are moving or what gravitational field we are immersed in? Einstein did more, perhaps, than anyone else before the advent of quantum mechanics to contribute to the queasy feeling that nothing is as it seems. Take this quotation from him about time: "I have realized that the past and future are real illusions, that they exist in the present, which is what there is and all there is." It's hard to imagine a more radical statement, and Einstein himself was uncomfortable with how unreliable our acceptance of the everyday world actually is—after all, to accept that the past and the future are illusions would disrupt a world that runs on the assumption that the passage of time is totally real.

IS EVERYTHING RELATIVE?

The year 2015 marked the hundredth anniversary of Einstein's final version of relativity, known as the General Theory of Relativity, and yet the most radical implications of it haven't sunk in, not as it concerns what is real and what is illusion. We are all used to accepting relativity in our everyday life, though we don't use that label. When your toddler draws on the wall with crayons, throws food on the floor, or wets the bed, you are much more likely to be indulgent about his behavior than if your neighbor's toddler comes to your house and does the same things. We are also used to the mind's fooling us about what our senses are detecting. Let's say you are going to a party and are told in advance that Mr. X, who will be there, is on trial for multiple burglaries in your area. At the party Mr. X comes up to you and casually asks, "Where do you live?" The sounds arriving in your brain through the mechanics of hearing will produce a very different response than if someone else had asked the same question.

Einstein could see in his mind's eye that objects would not appear to travel at the same speed to someone riding a beam of light and to someone standing on another moving object. Since the speed of anything is measured by the time it takes to travel a certain distance, suddenly time and space had to be relative as well. Very soon Einstein's chain of reasoning became complicated—it took ten years, from 1905 to 1915, for him to consult mathematicians in order to find the correct formulation of his theory. In the end, the General Theory of Relativity was hailed as the greatest piece of science ever accomplished by a single mind. But it mustn't be lost that Einstein cracked the code of space, time, matter, energy, and gravity by using the *experience* of visual images.

Does this prove that you are creating your own personal reality according to your own experiences? Of course. Every moment of the day you relate to reality through all kinds of filters that are uniquely your own. A person you love is disliked by someone else. A color you find beautiful is ugly to another person. A job interview that sends you into an immediate stress response poses no threat to a job applicant who happens to be more self-confident. The real question isn't whether you are creating reality—all of us are—but how deeply your interventions go. Is there anything that is real "out there" independent of us?

Our answer is no. Everything known to be real, from a subatomic particle to billions of galaxies, from the big bang to the possible end of the universe, is keyed to observation and as such to human beings. If something is real beyond our experience, we'll never know it. Let's make clear that we aren't taking a position that is nonscientific or anti-science. While Einstein was seeing images in his mind's eye that would overturn time and space, other pioneers in quantum physics were dismantling everyday reality even more radically. Whereas the theories of relativity were mostly the product of one person (with some help from colleagues), quantum physics was developed collectively by many physicists in Europe. Solid objects were now seen as energy clouds. The atom was observed to be mostly empty space (if a proton were the size of a grain of sand sitting in the center of a domed football stadium, an electron would be orbiting it at the height of the ceiling).

One by one, the quantum revolution that exploded in Einstein's lifetime took away every reliable bit of the world "out there." Intellectually, the consequences were devastating. There's a famous aphorism, uttered by astronomer and physicist Sir Arthur Eddington as he contemplated the peculiarities of the quantum domain: "Something unknown is doing we don't know what." These words are generally taken as a quip from a bygone era. Eddington, who offered some of the first proof that the theory of relativity actually matched reality, lived at a time before physics aimed its sights at a total explanation of the cosmos—a

Theory of Everything—which some believe is just around the corner.

But the quip (something Eddington had a knack for) should be taken seriously. Even a confident mind like Stephen Hawking's has more or less given up on a Theory of Everything, settling for a patchwork of smaller theories that will serve to explain how local aspects of reality work, not the whole. But can it really be true that reality is so mysterious that all of us have been mistaken about it since we were born?

THE QUANTUM AND THE APPLECART

Relativity was such a mind-bending theory that in the popular imagination, it seemed to go as far as physics could go. But this was far from the case. The story of what is real and what isn't took an uncomfortable turn known as the quantum revolution. This didn't happen totally independently of Einstein's work. A huge amount of knowledge is contained in $E = mc^2$, which applies to phenomena as diverse as black holes and splitting atoms. Yet, in a sense, the most startling aspect of $E = mc^2$ is the equal sign.

"Equal" means "the same as," and in this case, energy is the same as matter, or mass is equivalent to energy. As far as the five senses are concerned, a sand dune, a eucalyptus tree, and a loaf of bread (matter) are totally unlike a bolt of lightning, a rainbow, and the magnetism that moves a compass needle (energy). But Einstein's formulation has been proved correct many times over. The same cannot be said of the trouble that ensued from it. By portraying nature as endlessly transformable, with matter possibly turning into energy, $E = mc^2$, as in nuclear reactions, raised the question of how this behavior works.

It was realized, to the discomfort of anyone who trusts in the everyday world of sand dunes, trees, and rainbows, that the building blocks of nature, bits of energy or quanta, sometimes behave like energy and sometimes like particles. The most common example is light. When it acts like energy, light behaves like waves; these waves can be divided into wavelengths, which is why rainbows and prisms prove that the sun's white light is actually an amalgam of many separate colors, each with its own signature wavelength. But light, when behaving like matter, travels in particles (photons) that are discrete packets of energy. In Latin, the word for "how much" is *quantum*, and this was the name chosen by physicist Max Planck, who originated the quantum revolution in December 1900 and won the Nobel Prize in 1918. The term denotes the smallest amount or packet of energy.

If $E = mc^2$ implies that nature could in principle be reduced to a simple equation—something Einstein believed until the end of his life—his breakthrough with relativity was headed for a collision with quantum theory, whose equations are not compatible with General Relativity. This collision plagues physicists even today, and it caused a rupture in the story of what is real and what isn't. The difficulty doesn't sound earthshaking on the surface. It's simply about big things versus small things. All the big things in the world, from Newton's apple to far-flung galaxies, behave as Einstein's General Theory of Relativity say they should. But the smallest things, the quanta, or subatomic particles, obey a separate set of rules, which turn out to be quite bizarre, or spooky, to use Einstein's term.

We'll get into the details of this spooky behavior soon, but for the moment, the big picture is what matters. By the late 1920s, everyone agreed that relativity and quantum theory were incredibly successful in their own right, and everyone also agreed that the two didn't mesh. The hot-button issue was gravity and its incredible nonlinear (curved) effects. Einstein had revolutionized gravity by the use of visual images to pose new answers. Besides the image of a body in free fall, mentioned above, here's another. Einstein imagined a passenger standing in an elevator as it accelerated upward in a building. The passenger feels himself growing heavier, but since his perspective is confined to the inside of the elevator, he has no way of knowing why he is getting heavier. From his perspective, the cause could be a change in gravity or the result of accelerated motion. Both explanations work; therefore, Einstein reasoned, gravity doesn't have a privileged place as a force.

Instead, it must be included in the constant transformation of nature, only in this case what's changing isn't matter into energy and back again. Gravity changes from a constant force to the curvature of space and time, which varies from place to place. Imagine that you are walking across a flat expanse of snow on a winter day. Suddenly you slip and fall into a drainage ditch concealed under the snow. Quick as a wink, you slide down the curved side of the ditch. You would travel faster than on the flat snow, and your weight increases, as you'd discover when you came to a crashing bump at the bottom of the ditch. In the same way, space is curved around big objects like stars and planets. When light, traveling in a straight line, comes near such an object, Einstein theorized that gravity, through the curvature of space, would cause the light's path to bend. (The proof of his prediction in 1919 was tremendously exciting—we'll discuss it in a later chapter.)

In one stroke Einstein turned gravity from a force into a fact of space-time geometry. But on the quantum side of the enterprise, physicists still continue to refer to gravity as one of the four fundamental forces in nature. The other three forces—electromagnetism and the strong and weak nuclear forces—have been observed to behave like light, sometimes being wave-like, sometimes particle-like. But for decades no one could find gravitational waves or the gravity particle (already named the graviton). Therefore, the confirmation of gravitational waves in late 2015 was tremendously exciting news.

Einstein's General Theory of Relativity had predicted such

waves, amazingly enough, although no one at the time had any inkling of how they could be detected. Even with highly sophisticated modern technology, detecting gravitational waves seemed impossible because of their weakness. In simplest form, we can envision the big bang sending ripples through the fabric of space 13.7 billion years ago, and yet attempts to detect these waves always ran into problems. Background radiation causes interference, for one thing, which meant that pinpointing a gravitational wave would be roughly like dropping a pebble into a stormy sea and trying to single out the disturbance only it made.

Then a project named Laser Interferometer Gravitational-Wave Observatory (LIGO) was funded with the ambitious goal of building gigantic 2 kilometer measuring devices calibrated to within 1/1000 of the radius of the nucleus of an atom in order to capture the signals of gravitational waves from cosmic sources, which didn't have to be the big bang. Gravitational ripples could be caused, theoretically, by immense cataclysms in outer space.

A few days after LIGO went into operation, in September 2015, by coincidence the gravitational waves given off by the collision of two black holes 1.3 billion years ago passed through the Earth and were picked up. Such an event sends ripples through space-time traveling at the speed of light. The success of LIGO marks the beginning of a new way to measure the universe, because gravitational waves can pass through stars, revealing their core, which is hidden from sight. They may lead cosmologists back to the very early universe, providing new insights, such as the formation of black holes.

But in other ways gravitational waves are irrelevant to the larger situation that modern science finds itself in. They serve as a distraction from the unsolved mysteries that could actually shift the paradigm regarding how we see reality. For one thing, the confirmation of gravitational waves wasn't a surprise or breakthrough in terms of understanding the universe. They fulfilled a prediction that was almost a century old, and most physicists fully expected them to exist. The cosmos didn't gain a new phenomenon.

Most physicists will acknowledge that there is still a rift in the story of reality. As it happens, this rift leads to a remarkable possibility. Our minds, including the stream of everyday thoughts that run through our heads, might be influencing reality "out there." This could be why small things don't behave the way big things do. For example, visualize a lemon in your mind's eye. See its nubbly yellow surface and oily rind. Now see a knife cutting the lemon in half. Tiny droplets of lemon juice spritz out as the knife cuts through the lemon's pale flesh.

As you did this visualization, did you find yourself salivating? That's the predictable reaction, because simply seeing the mental image of a lemon creates the same physical response as an actual lemon. This is an example of an event "in here" causing an event "out there." The molecules that send a message from the brain to the salivary glands are no different from the molecules "out there" in lemons and rocks and trees. The body, after all, holds the same status as a physical object. We pull similar feats of mind over matter constantly. Every thought requires a physical change in the brain, all the way down to the activity of our genes. Microvolts of electricity fire along billions of neurons while chemical reactions take place across the synapses (or gaps) that separate brain cells. And the pattern of these events isn't automatic; it shifts according to how you experience the world.

Mind over matter upsets the applecart of physics through this discovery, that the act of observation—mere looking—isn't passive. If you look around the room you're sitting in at this moment, the things you observe—walls, furniture, light fixtures, books—don't get altered. Your gaze seems completely passive. But as far as what's going on "in here," no gaze is passive. You are altering the activity in your brain's visual cortex as your eye falls upon different objects. If you happen to see a mouse in the corner, a riot of activity may be set off in your brain. What we take for granted, however, is that seeing things is passive "out there." This is where the theory of quantum mechanics caused an upset.

If you move from big things to small things, observing photons, electrons, and other subatomic particles creates a mysterious phenomenon known as the observer effect. We already mentioned that photons and other elementary particles have a wave-like aspect and a particle-like aspect, but they can't have both at the same time. According to quantum theory, as long as a photon or electron isn't being observed, it acts like a wave. One feature of waves is that they spread out in all directions; there is no pinpoint location for a photon when it is in its wave-like state. Yet, as soon as the photon or electron is observed, it behaves like a particle, displaying a specific location along with other features like charge and momentum.

We will leave until later the specifics of complementarity and the uncertainty principle, two formulations that are critical for quantum behavior. The thing to concentrate on for now is the possibility that very tiny things "out there" can be altered simply through looking, which is a mental act. Common sense finds this hard to accept, because we're so used to assuming that gazing is a passive act. Go back to the mouse in the corner. When you happen to see a mouse, it often freezes and then quickly scurries away in an attempt to survive potential attack. Your gaze caused this reaction for the simple reason that the mouse sensed you looking at it. Can a photon or electron sense a scientist looking at it?

The very question sounds preposterous to scientists who maintain, as the vast majority do, that mind isn't present in nature, at least not until a series of happy accidents caused human life to evolve on earth. Nature is both random and mindless, according to a scientific credo assumed for centuries to be true. So how could a prominent contemporary physicist like Freeman Dyson say the following?

Atoms in the laboratory are weird stuff, behaving like active agents rather than inert substances. They make unpredictable choices between alternative possibilities according to the laws of quantum mechanics. It appears that mind, as manifested by the capacity to make choices, is to some extent inherent in every atom.

Dyson's statement is daring on two counts. He is claiming that atoms make choices, which is a sign of mind. He is also saying that the universe itself exhibits mind. In one stroke this bridges how big things and small things behave. Instead of atoms behaving totally differently from clouds, trees, elephants, and planets, they only *appear* different. If you look at dust motes dancing in a beam of sunlight, their motion appears totally random, which is how the physics of bodies in motion would describe them. But another visualization helps to make the situation clearer.

See yourself perched on the observation deck of the Empire State Building with a physicist next to you. You are both gazing down at the street below. At each corner some cars turn left and others right. Is this a random pattern? Yes, the physicist replies. A statistical array can be charted to show that over a period of time, just as many cars turn right as left. In addition, no one can reliably predict if the next car coming to a corner will turn right or left—the odds are 50/50. But you know that this is a case where appearances are deceiving. Every driver inside those cars has a reason for turning right or left; therefore, not a single turn is random at all. You just have to know the difference between choice and chance.

In science, the notion of chance is so dominant that mentioning the possibility of choice, as it pertains to physical objects, verges on the absurd. Consider our own planet: all the elements that are as heavy as iron or heavier—including many common metals and radioactive elements like uranium and plutonium—originated in the explosion of giant stars known as supernovas.

Without such explosions, even the incredible heat inside a regular star like our sun isn't enough to bond atoms into the heavier elements. Once a supernova explodes, these heavy elements become interstellar dust. The dust gathers into clouds, and in the case of our solar system, these clouds eventually coalesced into planets. The molten core of Earth is made of iron, but there are currents inside it that carry some of the iron close to the planet's surface. A bit of iron even leaches into the oceans and the upper layers of soil. From it, you got the iron that makes your blood red and allows you to breathe by picking up oxygen from the air.

Even though the floating dust motes in a beam of sunlight are exactly like the star dust that randomly floats among the galaxies, the fate of some star dust was unique. Some dust turned into a vital aspect of life on Earth. You, a human creature, act with purpose, meaning, direction, and intention—the very opposite of randomness. How did something random become something nonrandom? How did meaningless dust produce the human body, which is your vehicle for pursuing everything meaningful in our lives? The answer, if Freeman Dyson is right, is mind. If mind links little things and big things, then dividing the universe into random events and nonrandom events misses the point. The point is that mind may be everywhere, and our lives happen to reflect this fact.

A POET FINDS AN ESCAPE ROUTE

Because Einstein is almost the symbol of a staggeringly great mind, most people don't realize that after the great triumph of General Relativity, which took place when he was just in his mid-thirties, Einstein bet on the wrong side of modern physics, because he couldn't accept its conclusions. When he famously said that he didn't believe that God played dice with the universe, Einstein was stating his opposition to the uncertainty and randomness of quantum behavior. He placed his lifelong faith on a unified creation that operated without rifts, tears, and separations.

The notion that there is one reality and not two was something Einstein strove to prove until his death in 1955, but this quest was so far from the mainstream of physics that he was considered an incidental thinker after the 1930s—in their franker moments, even his greatest admirers shook their heads over such a great mind spending decades chasing a will-o'-the-wisp. But on one occasion he was given a clue about how to escape the trap posed by relativity and quantum mechanics. The escape route wasn't scientific, however, but came from a poet.

On July 14, 1930, reporters from around the world gathered outside Einstein's house in Caputh, a village outlying Berlin favored by the well-to-do as an escape from the hustle and bustle of the city. The occasion was a visit by Rabindranath Tagore, a great Indian poet then at the height of his fame. Born to a prominent Bengali family in 1861, almost twenty years before Einstein, Tagore had leapt into the Western imagination by winning the Nobel Prize for Literature in 1913. He was also a philosopher and musician, someone the West viewed as an embodiment of Indian spiritual traditions. The purpose of Tagore's visit with "the world's greatest scientist," as Einstein was popularly—and probably rightly—known, was to discuss the nature of reality.

As science raised serious doubts about the religious worldview, readers felt that Tagore enjoyed an uncanny and very personal connection with a higher world. Reading even a few snippets of his writings creates the same impression today.

I feel this pang inside –
Is it my soul trying to break out,
Or the world's soul trying to break in?

My mind trembles with the shimmering leaves.
My heart sings with the touch of sunlight.
My life is glad to be floating with all things
Into the blue of space and the dark of time.

On that July day, as their conversation was recorded for posterity, Einstein was more than politely curious about Tagore's worldview—he recognized the appeal of an alternative reality.

Einstein asked the first question. "Do you believe in the Divine as isolated from the world?"

Tagore's reply, delivered in flowery Indian English, was a surprise. "Not isolated. The infinite personality of man comprehends the universe. There cannot be anything that cannot be subsumed by the human personality . . . the truth of the Universe is human truth."

Tagore then set out a theme that blended science and mysticism in a metaphor. "Matter is composed of protons and electrons, with gaps between them, but matter may seem to be solid without the links in spaces which unify the individual electrons and protons . . . The entire universe is linked up with us, as individuals, in a similar manner—it is a human universe."

In a simple phrase—*the human universe*—Tagore had posed the ultimate challenge to materialism. He had also undermined the cherished belief in a divine universe. Materialism would place human beings as an accidental creation that occurred on a speck of a planet awash in billions of galaxies. Religion, in its most literal interpretation, would place God's mind infinitely beyond the human mind. Tagore believed neither of these things, and Einstein immediately became engaged, as the transcript shows.

EINSTEIN: There are two different conceptions about the nature of the universe—the world as a unity dependent on humanity, and the world as a reality independent of the human factor.

Tagore renounced this either/or proposition.

TAGORE: When our universe is in harmony with man the eternal, we know it as truth, we feel it as beauty.
EINSTEIN: This is the purely human conception of the universe.
TAGORE: There can be no other conception.

He wasn't spouting poetic fancy, or even mystical dogma. Tagore—flowing robes and sage's long white beard notwithstanding—for seventy years had been coming to terms with the scientific view of reality, and he felt that he could counter it with something deeper and closer to the truth.

TAGORE: This world is a human world . . . the world apart from us does not exist. It is a relative world, depending for its reality upon our consciousness.

No doubt Einstein understood the implications of Tagore's "human universe," and he didn't ridicule it or attempt to undermine it. But he couldn't accept it, either. The most pointed exchange immediately followed.

EINSTEIN: Truth, then, or beauty is not independent of man?
TAGORE: No.
EINSTEIN: If there would be no human beings anymore, the *Apollo Belvedere* [a famous classical statute in the Vatican] would no longer be beautiful.
TAGORE: No!
EINSTEIN: I agree with regard to this conception of beauty, but not with regard to truth.
TAGORE: Why not? Truth is realized through man.

EINSTEIN: I cannot prove that my conception is right, but that is my religion.

It was astonishingly modest for Einstein to say that he couldn't prove that truth is independent of human beings, which is of course the cornerstone of objective science. Human beings don't have to exist for water to be H_2O or for gravity to attract interstellar dust and form stars. By using the tactful word *religion*, Einstein said, in effect, "I have faith that the objective world is real, even though I can't prove it."

This once-famous meeting between two great minds is now largely forgotten. But in a startling way it was prophetic, because the possibility of a human universe, one that depends upon us for its very existence, now looms large. The most fantastic of possibilities, that we are the creators of reality, is no longer fantastic. After all, belief and disbelief are human creations, too.

PART ONE

THE ULTIMATE MYSTERIES

WHAT CAME BEFORE THE BIG BANG?

Though time and space had started to curve like a sagging clothesline, there wasn't wholesale panic in physics, because the chance that the line might snap apart didn't quite exist yet (black holes, which snap space and time, were brought into the picture later on). Brilliant equations are devised to keep reality intact, so the very fact that the mathematics was so arcane kept some very disturbing ideas away from the general public. But this all changed with the advent of the big bang theory. In one stroke, time snapped in two. There was time as we know it, which arrived on the scene with the big bang, and there was something else—weird time, pre-time, no time?—that existed outside our universe.

Let's see if we can visualize reality outside our universe. For the sake of convenience, we'll put the riddle this way: "What came before the big bang?" There's no better way to visualize the problem than stepping into an imaginary time machine that's whisking us back some 13.7 billion years. As we get close to the unimaginable explosion that began this universe's creation, our time machine is exposed to extreme danger. It took hundreds of thousands of years for the infant universe, which was superheated, to cool down enough for the first atoms to coalesce. But since our time machine is imaginary to begin with, we can imagine it coasting through superheated space without melting or flying apart into subatomic particles.

Getting within a few seconds of the big bang, or even less, we feel we're nearing the goal. "Seconds" means that time exists, and now the only challenge is to shave seconds down to millionths, billionths, and trillionths of a second. The human brain doesn't operate at such fine scales, but let's assume we have an onboard computer that can translate trillionths of a second into human terms. Eventually we arrive at the smallest unit of time (and space) that it is possible to imagine. William Blake's famous lines of verse, "Hold Infinity in the palm of your hand / And Eternity in an hour," is coming true, although an hour is much, much too long. At this point, when the scale of the cosmos was infinitesimally tiny, our onboard computer goes haywire and unexpectedly, nothing can compute.

Our whole frame of reference has dissolved. In the beginning, there was no matter of the kind we observe today, just a swirling chaos, and within this chaos, there may have been no rules of the kind we call the laws of nature. Without rules, time itself falls apart. The captain of our time machine turns to the passengers to tell them how bad the situation is, but unfortunately, he can't, for several reasons. As time collapses, so do concepts like "before" and "after." To the captain, we no longer left Earth at a certain time and arrived later at the big bang. Events are all gummed together in an unimaginable way. The passengers can't cry, "Let me out of here," either, because space has also dissolved, rendering "in" and "out" useless concepts.

This breakdown at the very threshold of creation is real, even if our time machine isn't. No matter how hard you work at it, regardless of how fine the slivers of time you shave, the threshold cannot be crossed—not by ordinary means, because, you see, the big bang "occurred everywhere," so it was not somewhere to where we could travel.

We are left with two options. Either "What came before the

big bang?" is an impossible question to answer, or else extraordinary means must be discovered that could possibly reveal an answer. One thing is certain, however: the origin of time and space didn't happen in time and space. It happened somewhere extraordinary, which, luckily for us, means that extraordinary answers aren't out of place—they are demanded. With that in mind, let the cosmic riddling begin.

GRASPING THE MYSTERY

"Before" and "after" are concepts that make sense only within the framework of space-time. You were born before you could walk; you will reach old age after middle age. The same isn't true of the birth of the universe. It has been widely theorized that time and space emerged with the big bang. If that's true—and it's only one possibility, not a fixed assumption—then the real question is "What came before time began?" Is that any better than the first way of putting it?

No. "Before time began" is a self-contradiction, like saying "when sugar wasn't sweet." We are squarely in the realm of impossible questions, but that's no reason to give up in advance. Quantum physics took to heart a conversation between Alice and the Red Queen in Lewis Carroll's *Through the Looking-Glass*. After Alice announces that she is seven and a half years old, the Queen retorts that she is a hundred and one, five months, and a day.

> "I can't believe that!" said Alice.
>
> "Can't you?" the Queen said in a pitying tone. "Try again: draw a long breath, and shut your eyes."
>
> Alice laughed. "There's no use trying," she said. "One can't believe impossible things."
>
> "I dare say you haven't had much practice," said the Queen. "When I was your age, I always did it for half an

hour a day. Why, sometimes I've believed as many as six impossible things before breakfast."

Quantum behavior forces us to be even more tolerant of impossible things. There is nothing ordinary about the conditions at the time of the big bang. To grasp them, some cherished beliefs must be challenged and then cast aside. First, one must realize that the big bang wasn't the beginning of the universe but of the *current* universe. Ignoring for now whether the current universe was created from another universe, physics can't actually trace the cosmos back to the absolute beginning. Taking measurements only works when you have something to measure, and in the very beginning there was an infinitesimal sliver of *something*, without order of any kind: no objects, no space-time continuum, no laws of nature. In other words, pure chaos. In this unimaginable state, all the matter and energy harnessed in hundreds of billions of galaxies was compressed. Within a fraction of a second, expansion accelerated with inconceivable speed. Inflation lasted between 10^{-36} (1/1 followed by 36 zeros) to about 10^{-32} seconds. By the time inflation ended, the universe had increased its size by a staggering factor of 10^{26}, while it cooled down by a factor of 100,000 times or so. A commonly accepted (but by no means definitive) scenario maps the birthing process as follows:

- 10^{-43} seconds—The big bang.
- 10^{-36} seconds—The universe undergoes a rapid expansion (known as cosmic inflation), under superheated conditions, enlarging from the size of an atom to the size of a grapefruit. There are no atoms in existence, however, or any light. In a state of near chaos, the constants and the laws of nature are thought to be in flux.
- 10^{-32} seconds—Still unimaginably hot, the universe boils with electrons, quarks, and other particles. The previous

rapid inflation decreases, or takes a pause, for reasons not fully understood.

- 10^{-6} seconds—Having cooled dramatically, the infant universe now gives rise to protons and neutrons that are formed from groups of quarks.
- 3 minutes—Charged particles exist but no atoms yet, and light cannot escape the dark fog that the universe has become.
- 300,000 years—The cooling process has reached a state where atoms of hydrogen and helium begin to form out of electrons, protons, and neutrons. Light can now escape, and how far it travels will determine from this point onward the outer edge (the event horizon) of the visible universe.
- 1 billion years—Through the attraction of gravity, hydrogen and helium coalesce into clouds that will give rise to stars and galaxies.

This time line follows the momentum produced by the big bang, which was sufficient, even when the universe was the size of a single atom, to produce, much later on, the billions of galaxies visible today. They continue to be driven apart by the expansion following the initial unimaginable primordial blast. Many complex events have occurred since the beginning (whole books are devoted to describing just the first three minutes of creation), but for our purposes, it's enough to view the rough outline.

Because we can all envision how a stick of dynamite or a volcano explodes, the big bang seems to fit our commonsense view of reality. But our grasp of what happened is fragile. In fact, the first seconds of creation call into question almost everything we perceive about time, space, matter, and energy. The great mystery about the emergence of our universe is how something was created out of nothing, and no one can truly comprehend how this occurred. On the one hand, "the nothing" is unreachable by any form of observation. On the other hand, the initial chaos of the infant universe is a totally alien state, being devoid of atoms, light, and perhaps even the four basic forces of nature.

This whole mystery can't be avoided, because the same birthing process continues, right this minute and all the time, at the subatomic level. Genesis is now. The subatomic particles that the cosmos is built upon wink in and out of existence continually. Like a cosmic on/off switch, there is a mechanism that turns nothing (the so-called vacuum state) into a teeming ocean of physical objects. Our commonsense view of reality sees the stars floating in a cold, empty void. In actuality, however, the void is rich with creative possibilities, which we see playing out all around us.

Already the argument feels like it's getting abstract, ready to float away like a helium balloon. We don't want that to happen. Every cosmic mystery has a human face. Imagine that you are sitting outside in a lawn chair on a summer day. A warm breeze makes you drowsy, and your mind is filled with half-seen images and half-realized thoughts. Suddenly someone asks, "What do you want for dinner?" You open your eyes and answer, "Lasagna." In this little scenario the mystery of the big bang is encapsulated. Your mind is capable of being empty, a blank. Chaotic images and thoughts roam across it. But when you are asked a question and make a reply, this emptiness comes to life. Out of infinite possibilities, you pick a single thought, and it forms in your mind of its own accord.

This last part is crucial. When you say "lasagna"—or any other word—you don't build it up from something smaller. You don't construct it at all; it just comes to you. For example, words can be broken down into letters, the way matter can be broken down into atoms. But of course, this isn't a true description of the creative process. All creation brings something out of nothing. It's humbling to realize that even as we feel comfortable being creators, immersed in infinite words and thoughts, we have no idea where they come from. Do you know your next thought? Even Einstein looked upon his most brilliant thoughts as happy accidents. The point is that creating something out of nothing is a human process, not a faraway cosmic event.

The transition of nothing into something always achieves the same result: a possibility becomes actual. Physics dehumanizes the process and does so with incredible precision. In unimaginably small scales of time, vibrations of quanta come out of emptiness and quickly merge back into emptiness, but this quantum on/off cycle is totally invisible to us. The rules governing physical creation must be deduced. You can't apply a stethoscope to the outside of the Superdome in order to discover the rules of football, and that's essentially what cosmology is doing, in attempting to explain the origin of the universe. Logical deduction is a great tool, but this may be a case in which it creates as many problems as it solves.

A BAFFLING BEGINNING

There's little doubt that the objects in space didn't exist before the big bang. But did space and time (technically, the space-time continuum) also emerge with them? The standard reply is yes. If there were once no objects, there was no space or time, either. So what was the pre-created state like? It didn't have an inside or outside, which are properties of space. As the infant universe expanded, it wasn't expanding with anything around it, and now, while billions of galaxies operate in outer space, the universe isn't like a balloon with a skin. Here again, the concepts of before and after, inside and outside simply don't apply.

Are we left with anything to hold on to? Barely. "To exist" suggests the possibility that even without time and space, things might happen. Here's a useful analogy. Imagine that you are sitting in a room where you notice that objects are moving slightly: the milk in your cereal bowl is jiggling, and you can feel a vibration coming up through the floor.

As it happens, you are deaf, so you have no way of knowing

if something is pounding on the walls of the room from the outside. (Some people might be sensitive enough to feel a vibration in their bodies—let's leave this aside.) But you can measure the waves in your cereal bowl and the vibrations of other objects, including the floor, ceiling, and walls. This is roughly how cosmologists confront the big bang. The universe is full of vibrations and waves emitted billions of years ago. These can be measured and inferences drawn from them. But uneasiness appears if we ask a simple question: Can someone who is deaf from birth actually know what sound is? Though there are measurable vibrations associated with sound, feeling them is not the same experience as hearing a solo violin, the voice of Ella Fitzgerald, or a dynamite explosion.

In the same way, measuring the light from racing galaxies and the background microwave radiation in the current universe (this radiation is a residue of the big bang) doesn't tell us what the beginning of the universe was like—we are working from inferences, just like a deaf person observing waves in his cereal bowl, and this limitation could be a fatal flaw in any explanation of where the universe came from.

We can still try, from our standpoint here in our space-time, to explore laws of nature that operate outside space and time. In particular, physics can resort to the language of mathematics in the hopes that its existence doesn't depend on which universe you happen to live in. Most of the speculation that follows keeps faith with mathematics as something eternally valid. Even in an alien universe, where time goes backward and people walk on the ceiling, if you add one apple to another apple, the sum is two apples, right?

However, no one has ever proved that this faith is actually valid. The mathematics that's applicable to black holes, for example, is locked in speculation, because a black hole is totally impenetrable. Mathematics could be the product of the human brain. Take the number zero. It hasn't always been around. By

1747 BCE, the ancient Egyptians and Babylonians had a written symbol for zero as a concept, but it wasn't used as a number for calculating purposes until around AD 800, in India, long past the heyday of Greek and Roman culture.

Zero means that nothing is there, and in mathematics, "nothing" is simply another number, not a sign of existential despair. "I've accomplished zero with my life" would be a despairing statement, but the equation 1–1 = 0 isn't. In quantum physics, concepts of time can be fiddled with in very peculiar ways without anyone's feeling distress over their own existence. If time started to behave peculiarly in the everyday world, that's a different story. Drifting between two worlds, something about time is mysteriously personal, and it must be explained if we want to understand a human universe.

THE BEST ANSWERS SO FAR

Clearly, the transition from early chaos to the orderliness of the current universe is filled with mysteries. The level where space and time break down is known as the Planck scale (named after German physicist Max Planck, the father of quantum mechanics), which is 20 orders of magnitude smaller than the nucleus of an atom (i.e., 1/10 followed by 20 zeros). Impressively, the presence of near chaos hasn't stymied human understanding. The mind still finds things that hold steady—perhaps.

The relevant measurements at such a small scale are still defined by three constants relating to very basic aspects of creation: gravity, electromagnetism, and quantum mechanics. During the Planck era, the incredibly minute timescale when the big bang began, Nature wasn't so recognizable, because the familiar constants and forces were either very different or didn't yet exist. In the so-called Planck dimension, space becomes "foamy," an indistinct state where any sense of direction, such as up and down, comes to an end. In terms of duration, Planck time—the characteristic scale of the Planck era—is more than 30 orders of magnitude faster than the fastest timescales of present-day nanoscience, where a nanosecond is a billionth of a second.

Therefore, the question of what existed before the big bang is equivalent to asking what existed before or beyond the Planck era. As it happens, physics can actually inquire about the trans-Planck realm. We know that mathematical laws govern the four basic forces: gravity, electromagnetism, and the strong and weak nuclear forces. This is one reason why faith in mathematics seems totally justified. Certain known constants tell us why these four forces assume the values they have in our universe. For example, when calculating the gravity anywhere—on Mars, on a distant star light-years away, or at the microscopic scale of atoms—no matter how different these environments are, the constant that applies to gravity remains the same. Relying on constants allows earthbound physics to travel mentally to the farthest reaches of space and time.

Could it be that the same constants exist in a timeless fashion, extending beyond our universe? Current physics cannot provide a definite answer. But if the constants are timeless, one can envision a continuity between our reality and unseen dimensions. Even short of that, you can see the allure of timeless constants. They give reality a sense of stability in the midst of roiling chaos. Timeless constants also shore up mathematics as a language that can survive the collapse of words. If "before" is a word that becomes meaningless, the value of pi (π) and the formula $E = mc^2$ will still hold up. But these, too, could be illusions when we cross the Planck-era threshold. For one thing, timeless constants beg the question of where they came from, leaving us without the origins story we are trying to seek.

Taking our inquiry as close as possible to the very beginning, one is tempted to identify the pre-created state as the quantum vacuum. In classical physics the vacuum is truly empty. Ironically, that kind of pure nothingness agrees with religious creation stories. ("And the earth was without form, and void; and darkness was upon the face of the deep"—Genesis 1:2.) But quantum theory and its derivatives state that the vacuum is not empty at all. It is filled with quantum "stuff." In fact, the quantum vacuum is as full as it can be, containing vast amounts of energy not manifesting in the observable universe. There is therefore no problem having the universe come out of the quantum vacuum, at least in terms of sufficient potential energies being available. There is also no doubt that following the universe to the very earliest phase must involve the physics of (quantum) vacuum. Even so, the Planck era sets up an impenetrable veil that blocks our view of the very beginning. One clever ploy is to do without a beginning at all, which has become a popular notion, strange as it sounds.

IS THE BIG BANG NECESSARY?

Theoretically, there are other possibilities besides the big bang. This sounds peculiar if the big bang is real. But remember, the explosion that began the universe wasn't like a dynamite explosion. There was no matter or energy of the kind that creation is now filled with. The visuals you see on TV science programs that resemble an exploding star embedded in the blackness of space are totally misleading, since there was no space in existence at the very beginning. It would make life easier if the universe were born another way.

A model called the steady state universe was proposed in 1948 by Hermann Bondi, Thomas Gold, and Fred Hoyle precisely to avoid the question of origin and what existed before the beginning. In the steady state model, the universe also expands forever, as in the big bang, but with the additional proviso that it always looks the same—it obeys the perfect cosmological principle, which means the universe looks the same everywhere and at all times. In other words, no matter where one looks, no matter how far back, the universe would be the same. This implies that the creation of matter takes place continually in space-time even as it is expanding.

According to the big bang theory, creation occurred once—nothing was required to turn into everything. So which model is true? Observations of distant light sources from the early state of the universe support an evolutionary model, which would discredit the original steady state. An updated version from 1993 proposed by Hoyle, Geoffrey Burbidge, and Jayant Narlikar, which they labeled as quasi-steady state, assumes that "mini bangs" repeatedly occur in the universe. Another alternative, known as chaotic inflation, is quite similar to steady state but at much, much larger scales. The term *chaotic inflation* was later replaced by *eternal inflation*, which gives a hint into its basic insight. Eternal inflation holds that certain "hot spots" in the quantum field accumulate enough energy to "pop" into creation, and this initial burst gives enough momentum that an entire universe can be born in an instant.

There are various reasons why eternal inflation has become much in vogue, but the main one is that onetime genesis can be turned into a constant behavior of the quantum vacuum. In essence, if the vacuum can bubble up with very tiny things (subatomic particles), why not give it the ability to bubble up with very large things (universes)? Inflationary theories all accept the big bang while also being saddled with the problem of beginnings (and endings). Eternity by definition has neither beginning nor ending. According to the principle of eternal inflation, space-time has always been bubbling up in various places with huge inflationary events, like a cosmic bubble bath. These events happen at the speed of light and continue forever.

Some brilliant physicists are infatuated with eternal inflation, and it's unlikely that someone as creaky and outdated as a philosopher could spoil things. But philosophy is concerned with words like *existence* and *eternity,* which turn out to be two very tricky terms.

GLIDING INTO THE MULTIVERSE

Eternal inflation ties in with another notion currently in vogue, the multiverse. In this scheme our universe isn't unique but only one of many, many universes—bubbles in the bubble bath—whose number could be nearly infinite (we'll go into this in detail later). Because the big bang is so widely accepted, the possibility of eternal inflation has a leg up on steady state theories. Once the door is opened, there are as many shots at creating a universe fit for human life as you may desire. In the cosmic casino, nature fizzes away with universes, and odds are it will hit upon the right one—our universe—eventually. After all, there are infinite rolls of the dice. The cosmic casino even allows for infinite changes in the rules (i.e., laws of nature) governing how a cosmos works. Gravity, the speed of light, the quantum itself can be jiggled as you please—so the theory goes.

But imagine that you are in a car with a friend acting as navigator. You're in unknown country, so you ask him which way to turn at the next intersection. He replies, "There are infinite ways to turn at the next intersection, but don't worry, they lead to infinite other intersections where we can also take an infinite number of turns. Eventually we'll get to Kansas City." Physics finds itself talking this way when dealing with the multiverse, eternal inflation, and the cosmic casino. The most absurd part, besides the fact that there are no data or experiments to show that a theoretical multiverse matches reality, is that such theories wave the map of infinite choices under our noses, claiming it's the best map anyone has ever drawn.

The standard view among cosmologists is that some combination of different models, perhaps including the quasi-steady state, may still be viable. But no matter how many universes are allowed, the theory still begs the question of what existed before the creative process began. *Before* remains a useless word, yet claiming that everything is, was, and always shall be the same feels intuitively like a hat trick.

There are other ways of avoiding the question of a beginning. Before the "big bang with cosmic inflation" model became established, many cosmologists had favored cycles of expansion and contraction leading from a beginning to an end and back again. In Eastern spiritual traditions, cyclical universes were accepted as a general concept taken from the life cycles of creatures being born, dying, and renewing themselves. Analogies aren't the same as scientific proof, but we need to remember that in the human universe, the processes that govern life as we know it must be tied to the mechanics of creation on a cosmic scale.

A variant of the cyclic universe would exclude a big bang popping out of nothingness while yet accounting for the present universe described by general relativity. Specifically, Roger Penrose has proposed a series of universes stretching back in infinite time. The current state emerged from a previous universe by recycling everything in it, and most important, the current physical laws and physical constants in nature. One big bang leads to another in an endless cycle, and so the pre-created state is just the tail end of the previous universe. The sequence of creations retains a certain kind of memory from one cycle to the next. In Penrose's intriguing conception, the entropy (or disorder) found in the universe plays a fundamental role. There is a law in physics (the second law of thermodynamics) that holds that the disorder in the entire universe increases over time. The words sound abstract, but it's this law that governed how a superheated early universe grew cold, how stars die, and why a log put on the fireplace goes up in smoke and leaves behind ashes. On scales both large and small, entropy increases.

There are islands of negative entropy in the universe, where energy can be used for more order, as in living ecosystems, instead of winding down or dissipating. You are an island of order. As long as you keep consuming food, air, and water, your body is such an island, turning raw energy into orderly processes in trillions of cells, renewing and replenishing them. Earth became an island of negative entropy, on the surface at least, when photosynthesis began billions of years ago. Plants convert sunlight into orderly processes, just as your body does. Turning into an energy consumer instead of an energy loser is critical. Disorder causes energy to dissipate into heat, like the heat given off by a bonfire. To combat this entropy, living creatures consume the extra energy needed to counter the loss. A fallen tree in the woods has lost the ability to get energy from the sun, and therefore disintegration and decay begin to do their work.

Penrose didn't argue against the second law of thermodynamics—he acknowledged that the entire cosmos is becoming colder, more spread out, and more disorderly. His objection specifically targeted inflationary theories of the cosmos. If disorder increases as time passes, he pointed out, then the reverse must be true—if you go back in time, any system will display more orderliness early on. For example, if you reverse time, the smoke and ashes given off by a bonfire would re-form into a piece of wood, and a rotting tree would return to being alive and growing. Therefore, the early universe should be the most orderly state of all—yet it wasn't. The Planck era was a time of pure chaos. So where did the "specialness" (Penrose's term) of the universe come from, making possible the development of life on Earth? Nothing about the early universe, from its first instant of utter chaos, seems to prepare the way for the evolution of galaxies so that life on this planet is favored in advance.

To a layperson, Penrose's objection to inflationary theories seems entirely cogent, although there are technical considerations brought up by skeptical cosmologists. He makes a second point that is subtler. Let's say we accept that life on Earth is so unique that the early universe had to pave the way through special conditions. Let's even accept that there were special conditions emerging when the cosmos was superheated and infinitesimally small. What about the rest of the vast universe? Life evolved on our planet regardless of what was happening in billions of other galaxies—we didn't need them. So how could the universe be set up to aid our evolution, if that is indeed true, while everywhere else doesn't look special at all? It's far more likely, Penrose declares, that the conditions for life on Earth became special later on. Perhaps it was only a matter of random chance. The less improbable explanation is the one science must choose.

Recently astronomers have somewhat undercut Penrose's objection with the discovery of thousands of stars with planetary systems. Some of these stars are enough like the sun that they could foster life on planets similar to life on this one. Great excitement followed the news that we are probably not alone in the universe. However, the good mood fades when it's pointed out that "probably" doesn't actually explain how life evolved from lifeless chemicals. The odds could be so long—millions and millions to one—that even a multitude of suns in faraway galaxies aren't enough to find the magic key to life. The objection can't be refuted; on the other hand, it can't be proven either. But as soon as you start talking about odds and probabilities, you are assuming that life evolved randomly, and "specialness" has taken a severe blow.

AN INGENIOUS INFORMATION THEORY

Or maybe not. When a theory has been as successful as the big bang in explaining how the universe evolved, posing objections is tricky. You may simply be pointing out glitches that can be patched up. Surely it would take a lethal blow to knock down the entire structure built up so carefully since the 1970s. But Penrose's argument about the second law of thermodynamics is so basic that it could topple the whole house of cards. The problem of cosmic inflation and that it did not emerge on its own as a natural evolution in scientific theories but rather was put together to account for some baffling mysteries of the older big bang cosmology. Inflation is well-supported by sensitive measurements. Its main thrust is to rescue the early universe from apparent chaos, but we need a source of orderliness that is more sophisticated than a bingo machine tossing out numbers at random.

Noted American cosmologist Lee Smolin has proposed some intriguing ideas about the geometry of the Planck era that could save it from pure chaos. Perhaps something immaterial was the source of orderliness, even if there was only chaos at the physical level during that time. Penrose and Smolin nominate information as the key ingredient. This seems like an intriguing thread to follow, because other physicists have theorized that when all matter and energy is sucked into a black hole and annihilated, information manages to survive. Proving this is very difficult, or even impossible, since the interior of a black hole is impenetrable, but it's an intriguing way to sidestep the "heat death" of entropy. What if information isn't disturbed even under the most extreme physical conditions? Ones and zeros can't freeze to death or be reduced to ashes in a fire. Perhaps the pre-created state was rich with information that was immune to the second law applying at the moment of the big bang.

By analogy, the information you carry around in your mind can survive all kinds of physical threats. One piece of information is your name. Once you know your name, it doesn't matter if you travel to the steaming tropics or the South Pole, as heat and cold don't cause your name to freeze or boil over. Your name isn't affected if you descend to the bottom of Death Valley or climb Mount Everest. Generally, only death or extreme brain trauma could deprive us of this intimate bit of information. The same is true for much more complex things, since the storage capacity of the human mind is vast. (And in some rare cases, people have awakened from deep comas lasting years, recovered memories, and resumed their lives.)

The survival of information in humans makes a cyclic universe seem like a real possibility. If a previous universe gave birth to ours, perhaps the constants and the laws of nature could be passed on in the form of information, especially mathematical, since some fundamental mathematics must be involved, yet this way of thinking avoids calling mathematics a physical property. In Smolin's model, passing the cosmic baton occurs when new "eons" emerge from black hole singularities. An eon would be a cosmic unit of time; a singularity is the tiny speck left when everything has been sucked into a black hole. Theoretically, such a speck is singular because it hasn't disgorged the things that create differences—space, time, matter, and energy. (There's no solid evidence that singularities actually exist, even though they are mathematically plausible.) The notion is that the universe will ultimately collapse into a single point (a singularity), into which matter, energy, the forces of nature, and space-time vanish, only to reemerge through a new singularity.

In other words, the big bang was preceded by the big crunch. We don't know enough about black holes to say how information could survive them when nothing else does, and singularity remains a theoretical construct only. As it stands, then, claiming that information wasn't destroyed in the early cosmic cauldron seems like another hat trick. One way or another, whatever is happening inside a black hole is just as unreachable as the Planck era at the beginning of the universe. The same impenetrable wall blocks our sight.

TWANGING THE SUPERSTRING

Though many people are terrified by higher mathematics, it helps to realize that everything about reality that gets formulated into mathematics also exists as a concept. If you grasp the concept, you often go straight to the heart of what the math is trying to say. Math is really a condensed, universal language that allows descriptions of so-called physical processes, or, better, descriptions of our interactions with nature. Certainly no amount of higher math can redeem a false idea. In the debate between models that include a big bang and those that don't, the pros and cons are not easy to weigh. If math is the only thing cosmology can still rely upon, why not put the whole burden on it? Perhaps the only secure way to describe the pre-created state is to describe it as a reality where only pure mathematics can guide us. Or, to go a step further, perhaps the pre-created state consists of numbers and nothing else. This sounds like a strange proposition, yet some theories are willing to go there.

The leading example is string theory, which later morphed into superstring theory as its ambitions expanded. String theory arose to resolve some critical but arcane quantum issues and yet has broader implications for the mystery of how elementary particles like photons, quarks, and electrons can act like both particles and waves. Many physicists have dubbed this the central problem of quantum mechanics. A particle is like a tennis ball flying over the net; a wave is like the swirling air it leaves in its wake. They don't look at all alike. However, if a tennis ball and swirling air can be reduced to one common trait, this might solve the problem.

String theory says that the common trait is vibrations. Imagine a violin string vibrating to produce musical notes. The exact note is determined by where the violinist places a finger on the string. In similar fashion, string theory considers waves to be the vibration of an invisible string, with particles being the specific "notes" that appear in space-time. The analogy to music is a powerful one, in that subatomic "harmonies" (vibrations that resonate with each other) are thought to determine how quarks, bosons like photons and gravitons, and other specific particles relate to one another and build up complex structures. Just as the twelve notes of the Western musical scale turned into countless symphonies and other musical compositions and there is virtually no end to the possible permutations of those twelve notes, likewise, a few kinds of vibrating strings could be the basis for the proliferation of subatomic particles being discovered in high-speed particle accelerators.

Though skeptics like to point out that strings vibrating below the level of observable reality might be figments of the imagination, the appeal of string theory is that it refers back to pure mathematics. An advanced model, known as superstring theory, expanded the complexity of the necessary equations. At first there were five superstring models that appeared to be different, but in the mid-1990s they were shown to have subtle and complex similarities. What emerged as the pinnacle of mathematical modeling was M-theory, where the M can stand, as its chief creator Edward Witten has whimsically said, for "magic," "mystery," or "membrane."

Magic and mystery come into the picture because M-theory has no foundation in any experiment or observation. It pulls a mathematical rabbit out of the hat by harmonizing previous string-type theories, which themselves were not founded on experiments or observations. The fact that M-theory does such a good job—on paper—seems both magical and mysterious. The ultimate trick would be to show that the universe actually works the way it does on paper, and no one has even remotely pulled that one off. (The third M, membrane, is a technical term in physics to describe how certain quantum objects extend through space like sheets or vibrating membranes. Here we teeter on the edge of very complex equations that can be grasped only through higher mathematics, but it's possible to give you a conceptual framework.)

WHERE DID EVERYTHING GO?

How did reality turn so enigmatic that it had to be reduced to numbers? Physics is about physicality, but as we saw, physicality vanished in the quantum revolution. We are speaking of simple, basic physicality, the kind that the five senses allow us to experience when someone kicks a rock and finds it hard. Subtle physicality remained, in the form of the subatomic particles and waves that quantum physics deals in. But two related hurdles could not be overcome.

The first hurdle, which we touched on earlier, has to do with the incompatibility of big and small objects. Einstein's General Theory of Relativity does a magnificent job with large objects such as planets, stars, and galaxies and the universe itself. Through its understanding of gravity and the curvature of space-time, relativity is accepted as providing the deepest understanding of anything macroscopic, and of the large scale of the universe itself. At the opposite extreme, quantum mechanics (QM) has been just as successful describing the tiniest objects in nature, particularly subatomic particles. And from the beginning of their formulation, general relativity and QM have not meshed. Each makes accurate predictions within its own domain; experiments can be run and observations made. But finding a link between the biggest and smallest objects in the universe has been extremely difficult.

The second hurdle grew out of this dilemma. Once it was established that there are four fundamental forces in nature, consisting of gravity, electromagnetism, and the strong and weak nuclear forces, the possibility of uniting them into one unified theory presented itself. By the late seventies, with the discovery of quarks, the so-called standard model emerged that united the quantum world on three fronts. The force responsible for light, magnetism, and electricity (electromagnetism) was united with the two forces that hold atoms together (the strong and weak nuclear forces). A world of tiny objects had surrendered to mathematical conformity. This step was known as the standard model, and considering how many brilliant minds contributed to it, unifying the three fundamental forces deserves to be called grand.

Only gravity remained to complete this "theory of almost everything" (the nearest we might hope to get to the Holy Grail, a Theory of Everything). By analogy, imagine that someone is assembling a jigsaw puzzle of the Statue of Liberty. All the pieces are in place but the torch. That piece isn't in the box, so a search begins to find it. "Don't worry," we're told, "it's just one piece. Once we locate it, the whole picture will be complete. We're nearly there." Yet no matter how hard everyone searches, the missing piece can't be located. And, to everyone's dismay, when they go back to the puzzle, the Statue of Liberty is only a vague outline surrounded by dense fog.

Modern physics is divided into two camps. One believes that the picture of the universe is nearly complete, lacking only one piece, which will be discovered in the future as long as the search is persistent. The other camp believes that the missing piece makes the whole picture vague and doubtful. We could also call these the business-as-usual camp (build the biggest accelerator, create more powerful telescopes, do more calculations, spend more money) versus the revolutionary camp (start all over with a new model of the universe). Because the business-as-usual camp considers itself practical and pragmatic, its mantra is "Shut up and calculate," meaning that too much theory is nothing but idle speculation.

In order for the business-as-usual camp to be victorious in the end, it must pry some very stubbornly embedded particles from the quantum fabric; only then will its calculations be validated. So far, optimism runs high, ever since one of the most important of these particles, the Higgs boson, was finally observed in 2012. We've mentioned how the quantum vacuum bubbles up with subatomic particles. Some of these are so elusive that dislodging them requires enormous amounts of machinery in the form of large and expensive accelerators. By bombarding an atom at ultrahigh energy, the quantum vacuum sometimes pops out a new kind of particle. It's precise, painstaking work, but these new particles predicted in the next generation theories prove whether existing theories are actually correct. The Higgs boson was predicted to exist, and therefore its discovery, when confirmed, would indicate that the standard model matches reality. But the standard model isn't the end; it is not grand unification.

The function of the Higgs boson is to give mass to other fluctuations in the quantum field, a technicality we don't need to dwell on. But this function is basic to the existence of all created physical objects. The media latched on to the nickname of "God particle," which embarrasses almost all physicists. To them, the validation of the Higgs boson was a triumph because it fills out one of the last remaining fundamental particles—the torch of the Statue of Liberty has been found, and the theoretical picture is very nearly complete. Searching for the last missing piece took five decades, ever since British physicist Peter Higgs and others first proposed that the so-called Higgs field existed.

The new discovery fits a familiar pattern. The history of modern physics has been a triumphant parade of proven results that mesh with theoretical predictions. The Higgs boson may be an important link to how the four fundamental forces are connected, but it could also be the end of the parade, since bringing gravity into the fold may be impossible in terms of validation. The graviton, a theoretical particle that pops out of the field of gravity when it is excited, is far from being observed or observable. One obstacle is a matter of technology. By some estimates, an accelerator that might produce the acceleration and energy necessary to get us any closer to the origin of physical reality would have to be bigger than the circumference of the earth.

This obstacle doesn't have to end the story, though. Math can get around practical difficulties. There isn't a scale big enough to weigh a blue whale, but its weight can be determined using calculations about its size, the density of its mass, and comparisons with smaller whales and dolphins that can be weighed. But the business-as-usual camp finds itself waist deep in a mathematical swamp, while string theory, superstring theory, and M-theory have added layer upon layer of complexity, but nothing verifiable in real life.

It's strange that a failure to wriggle out of a very basic difficulty should call the entire cosmos into question. But reality is one thing, not two. The smallest and the largest things must be connected in some way. The fact that the connections are invisible doesn't stop mathematics. But the mathematics is so complicated, with large gaps remaining and obvious patches applied to the bald spots, adding to the impression that if one gets too far away from reality, even mathematics can't come to the rescue. Unless of course we admit that the unreasonable power of mathematics, as physicists say, is pointing to the mental nature, from which mathematics originates, of the cosmos.

WHY DOES THE UNIVERSE FIT TOGETHER SO PERFECTLY?

We say that the universe began with a bang, but actually the early universe was more like a shy performer emerging from its dressing room—the early universe took its time until every seam and stitch fit together perfectly. Billions of years later, we look around and are amazed that we inhabit a cosmos that fits human life perfectly—in fact, too perfectly. There is no reasonable way to explain how the big bang can have every seam and stitch in place. It's as if Leonardo da Vinci managed to paint *The Last Supper* by throwing paint randomly against a wall and hoping for the best.

Yet current cosmology insists that the early universe had to develop through random chance. There was no designer and certainly no designer behind the scenes. Scientific creation stories all exclude God in any form. But how do you get the incredible orderliness of human DNA, with its 3 billion basic chemical units, starting with a stick of cosmic dynamite? How does order come out of chaos, in other words?

An answer can't be found without using a good deal of brain power, and yet your brain is a perfect example of how the problem comes home in everyone's life. In order for you to read the words on this page, extremely precise processes must take place in the brain's visual cortex. The specks of ink on the page must register as meaningful information; the information has to be presented in a language you understand; as your eye passes from one word to the next, the meaning of each word is connected to that of the next word and then disappears out of view but not out of our minds.

This is miraculous enough, but the real mystery is that the molecules inside each brain cell are locked into fixed, predetermined actions and reactions. If you place iron in contact with free atoms of oxygen, they form ferrous oxide, or rust—every time. The atoms have no choice in the matter. They can't form salt or sugar instead. Meanwhile, despite the fixed laws of chemistry in the brain, you manage to have thousands of new experiences every day, jumbled in unique ways that make today different from yesterday or tomorrow.

So the evidence of the brain tells us that chaos and order won't necessarily have a simpler relationship. Chemistry is completely predetermined; thinking is free. If we can solve how they relate, the universe may yield up its deepest secret of all. More important, we'll discover how the mind works, which, to be candid, is more interesting to most people than the big bang.

GRASPING THE MYSTERY

In physics the riddle of why a random universe fits together so well is known as the fine-tuning problem. But before jumping into science, we can find clues in something much older—creation myths. And, though every culture has its own creation myths that arose and were passed down over many centuries, all the stories can be broken down into two classes. The first class explains creation through a familiar action that people can relate to. For example, in India, one myth says that the forces of light and darkness created the world by using a mountain, Mount Meru, like the paddle in a milk churn, pulling the paddle back and forth until butter solidified out of an ocean of milk.

The second class of myths wraps creation in a mystery by doing the exact opposite, trying to show that the world was created by totally supernatural means. The Judeo-Christian creation story in the book of Genesis adheres to this pattern: Yahweh begins with a void and magically turns it into light, the heavens, the earth, and all the creatures upon the earth. There is no similarity to anything in everyday life like churning butter—until now. Modern cosmology parallels Genesis by positing that the universe arose when something emerged from nothing. It would offend the scientific mind to call it magical or supernatural. So let's call it mysterious, which would be the understatement of the century.

Creation is very big. The universe appears potentially to extend 46 billion light-years as far as the eye, or the telescope, can see. This is how far light has traveled since the big bang. As the baby universe expanded, it didn't fly apart haphazardly. It began to take shape according to certain rules known as the constants of nature, rules that can be formulated with mathematical precision. A few of these constants have already appeared in the book, namely the speed of light and the constant of gravity.

Constants create order in nature, like old-fashioned mothers who saw it as their duty to have dinner on the table at the same time every night. The problem is, order and pattern had to come from somewhere, and the only somewhere anybody can prove is the big bang, and that was totally chaotic until suddenly it wasn't. Clearly something more is needed besides waiting around, and the same holds true for the universe—but what?

The physics community accepts that fine-tuning exists. Too much or too little gravity, too much or too little mass, too much or too little electric charge, would have caused the newborn universe either to collapse in on itself or to fly apart too fast for atoms and molecules to form. Therefore, stable stars could not have formed, or any of the complex structures in cosmic evolution. Further down the line, life on Earth would have been impossible without a variety of cosmic coincidences, such as the presence of the essential amino acids, the building blocks of proteins, which apparently existed in interstellar dust.

Physicists also agree that we have to discover where the constants of nature came from. Precise mathematical laws govern the four fundamental forces, gravity, electromagnetism, and the strong and weak nuclear forces. For example, when measuring the gravity at remotely separate locations, such as on Mars, or on a distant star light-years away, no matter how unlike these environments are, the constant that applies to gravity remains the same. Relying on constants allows earthbound physicists to travel mentally to the farthest reaches of space and time.

When they do this, some startling coincidences crop up. For example, out in deep space, the explosion of the very largest stars, massive supernovas, are occurrences that can be observed through powerful telescopes on Earth or orbiting around it. Supernova explosions that occurred billions of years ago are responsible for forming all the heavy elements in existence, such as calcium, phosphorus, iron, cobalt, and nickel, to name just a few. Atoms of these elements first circulated as interstellar dust, gravity caused them to clump together, and eventually they wound up inside the ancient solar nebula, where all planets, including our own, formed. The iron that makes your blood red came from a supernova that self-destructed eons ago. The specifics of the explosion are determined by the weak and strong forces, which exist at the infinitesimally small scale of the atomic nucleus. If these forces were different by as little as 1 percent or so, there would be no supernova explosions, no formation of heavy elements, and therefore no life as we know it. A particular constant governing the weak force had to be exactly what it turns out to be.

Let's consider some specific cases of fine-tuning at the level of everyday reality, where matter is comfortably composed of atoms and molecules. What's known as the fine structure constant determines the properties of these atoms and molecules. It is a pure number, approximately 1/137. If the fine structure constant were different by as little as around 1 percent, no atoms or molecules as we know them would exist. As relates to life on Earth, the fine structure constant determines how solar radiation is absorbed in our atmosphere, and it also applies to how photosynthesis works in plants.

The sun just happens to emit the majority of its radiation in a part of the spectrum where the atmosphere of Earth just happens to allow sunlight through without absorbing or deflecting it. Here we run into another perfect match between two extremes of nature. In this case, the perfect match allows for just the right portion of the spectrum to reach Earth's surface for plants to feed on. The gravitational constant (which governs the sun's radiation) is a macroscopic value, while the atmospheric transmission of sunlight, with only some wavelengths making it through, is determined by the fine structure constant and is applicable on a microscopic scale.

There is no clear reason why two constants, separately governing very big things and very small things, should mesh. (It's like discovering that a child's fingerprints can tell you if he'll grow up to be a brain surgeon.) Yet if these two effects didn't mesh together perfectly, there would be no life as we know it. With good reason the fine-tuning problem has been called one of the biggest embarrassments of physics, although biology can also take a share. Life depends on a fragile balance of constants, too. In fact, it was the total improbability of a universe that leads to life on Earth that brought fine-tuning into high relief. The existence of DNA involves too many coincidences, going back to the big bang itself. Theorists began to consider whether these coincidences are actually something else, an indicator that some deep underlying unity has been missed. The clues to this hidden unity are the suspiciously fine-tuned constants, although many other kinds of coincidences arouse the same suspicion.

Figuring out why the universe is so fine-tuned has preoccupied many cosmologists, and one contingent has long been uncomfortable assigning the universe to pure chance. Here's a famous passage from astronomer Fred Hoyle:

> *A junkyard contains all the bits and pieces of a Boeing 747, dismembered and in disarray. A whirlwind happens to blow through the yard. What is the chance that after its passage a fully assembled 747, ready to fly, will be found standing there? So small as to be negligible, even if a tornado were to blow through enough junkyards to fill the whole Universe.*

For the majority of working physicists, Hoyle's analogy doesn't hold water, because the equations underlying QM and its tremendous predictive power dictate the operation of random chance and uncertainty. Still, explaining why the constants are so fine-tuned defies current knowledge, and there's even the intriguing possibility that they must be fine-tuned in order for human beings to exist. What if chance had nothing to do with it?

THE BEST ANSWERS SO FAR

An attempt has been made to explain the why of fine-tuning through the *anthropic principle.* The term first appeared in 1972 at a conference celebrating the five-hundredth anniversary of the birth of Copernicus; the name is derived from *anthropos,* the Greek word for "(hu)man." The relevance of Copernicus is that a planetary system where the earth revolves around the sun takes away the central position of human beings in creation. One of the main authors of the anthropic principle, astrophysicist Brandon Carter, declared, "Although our situation is not necessarily central, it is inevitably privileged to some extent." His assertion was either a breakthrough or an outrage, depending on your beliefs. Returning human beings to a privileged place in a cosmos billions of light-years in size was bold, if nothing else. For a calm description of what the anthropic principle implies, we turn again to physicist and mathematician Sir Roger Penrose.

In his much-respected book *The Emperor's New Mind: Concerning Computers, Minds, and the Laws of Physics* (1989), Penrose says that the argument for giving human beings a privileged position is useful "to explain why the conditions happen to be just right for the existence of (intelligent) life on the Earth at the present time." Despite the allegiance of physics to randomness, Penrose points to "striking numerical relations that are observed to hold between the physical constants (the gravitational constant, the mass of the proton, the age of the universe, etc.). A puzzling aspect of this was that some of the relations hold only at the present epoch in Earth's history, so we appear, coincidentally, to be living at a very special time (give or take a few million years!)."

Being here, we look around and find that the cosmos led to our existence. A calm tone is necessary at this point, because on the fringes of the discussion are creationists who read the Bible literally, ready to pounce, claiming that physics now supports their belief that God gave man dominion over the earth, exactly as the book of Genesis teaches. Any such suggestion that human beings are divinely favored in the evolution of the cosmos is scientific heresy. But the anthropic principle isn't about having a religious agenda. It works from a remarkable fact that's hard to explain: intelligent life now exists on earth, namely us, and we are capable of measuring the constants that gave rise to intelligent life. Is this more than a coincidence?

An analogy may help. Imagine that jellyfish are intelligent and want to know what the ocean is made of. Jellyfish scientists analyze the ocean's chemical composition, and they make a surprising observation. "The chemicals inside our bodies exactly match the chemicals in seawater. The match is too perfect to be just a coincidence. There must be another explanation." They'd be right, because the reason that seawater and the liquid inside a jellyfish match is that evolution made it so—jellyfish wouldn't be alive without the sea.

DO HUMANS MATTER THAT MUCH?

The anthropic principle gained support among scientists who felt uncomfortable with coincidence piling upon coincidence, yet it gives us no definitive explanations that fit current science. As with jellyfish, it could be that evolution created a match between the human brain and the constants in the universe. Or not. They might match for some other reason, or the seeming match could be illusory and we will discover important kinds of mismatches if we keep looking. There are large areas of controversy about how accidental anything in the cosmos actually is, but at least the ice has been broken—the total lock on randomness has been broken intellectually. (The recent discoveries of planets orbiting around distant sunlike stars has boosted randomness, the notion being that there may be millions upon millions of planets potentially capable of sustaining life. If so, then Earth would be lucky in the cosmic lottery but not unique or perhaps all that special. Copernicus may have the last laugh.)

To bolster its credibility, the anthropic principle has been expressed in strong and weak versions. The *weak anthropic principle* (WAP) tries to take any special dispensation out of the equation. It makes no claim that intelligent life on Earth was somehow the goal of cosmic evolution starting with the big bang. WAP only says that the universe, if it is ever fully explained, must conform to life on Earth. Maybe the constants that we have been measuring contain some kind of wiggle room, so that our knowledge, while correct, is limited to our perspective. Imagine a bee that can only collect pollen from pink flowers. The weak bee principle would say that no matter how you talk about the evolution of flowers, a connection must be made between the pink ones and bees. The fact that there are lots of other flowers in other colors can be explained any way you want without worrying about bees.

The *strong anthropic principle* (SAP) makes a bolder claim: that there can be no knowable universe without human beings in it. The evolution of the cosmos must necessarily lead to us. Many physicists squirm at this suggestion, which smacks of metaphysics. One mischievous commentator went one step further with a so-called *very, very strong anthropic principle,* which he stated as "The universe came into existence so that I, personally, could argue cause-and-effect on this web page, specifically." This might seem like a joke, taking SAP to a ridiculous extreme. But, if the universe must accommodate human beings, there is no logical reason why it can't accommodate this very moment in time. Cause and effect doesn't have a mind of its own. If the constants lead to deterministic outcomes (e.g., dropping a ball always leads to the ball's falling to earth), it's just as easy for a moment in time—pick any one you want—to be predetermined.

Now you can see why a belief in cause and effect is one of the core beliefs that have broken down in the postquantum era. It just won't do to say that the big bang inevitably led to this very moment, the page you are now reading, the ham sandwich or cup of tea at your elbow, and the spelling of your last name. Strict cause and effect would mean that your next thought or the next word out of your mouth was predetermined 13.7 billion years ago. By turning strict cause and effect into probabilities, quantum mechanics eased this difficulty. We now live with "soft" cause and effect, you might say. Every event emerges from a set of probabilities, not an ironclad chain reaction.

Still, the mystery of the fine-tuned universe hasn't gone away. Probabilities can tell you the likelihood of an electron appearing at point A in time and space. It has nothing to say about how electrons came into existence as part of a fine-tuned universe. By analogy, if you have a friend with a vocabulary of 30,000 words, and you also know how often he uses each word, you can use probability to calculate the likelihood that his next word will be "jazz." Maybe he's not a jazz enthusiast, so the likelihood is very small, with a probability of 1 in 1,867,054. That's a powerful degree of precision. But you still have no way of explaining why he chose that word anytime that "jazz" escapes his lips. On a large scale, your skill at probabilities can't explain why language came into existence among primitive societies hundreds of thousands of years ago.

No matter whether the anthropic principle is weak or strong, it allows Earth to stop being a random speck afloat in the cosmic ocean. It's hard to get past the proposition that the constants of nature have their particular values because the universe is built to allow for life to develop. If you've ever idled away an afternoon building a house of cards, you know that the slightest slip in one card leads to the collapse of the whole structure. Imagine, instead of a house with fifty-two cards, that you are building human DNA, which has 3 billion base pairs, the chemical rungs spaced along the twisted ladder of the double helix.

Consider that the process for constructing human DNA took some 3.7 billion years from the first prototypes of life on Earth, and 10 billion years of cosmic existence to get to that point. How many slips could randomly occur along the way, causing DNA's house of cards to collapse? Too many to calculate. Your genes were inherited from your parents, but in the process of transmitting them, about 3 million irregularities, in the form of mutations, occurred on average. These random alterations in DNA, along with mutations caused by X-rays, cosmic rays, and other aspects of the environment, cast huge doubts about life as an accidental creation.

The rate of random mutations is statistically verifiable. In fact, this is the main way that we can trace where human genes traveled after the first band of our human ancestors migrated out of Africa 200,000 years ago. The mutations in their DNA serve as a kind of clock with which we can trace their path. So, randomness has powerful arguments in its favor, while at the same time statistics also undermine randomness, given how often DNA could have lost its way over 3.7 billion years. Yet all these slips were avoided, and this fact muddies the waters if you want to turn randomness into the only force at work. Life is poised on the cusp of order and disorder. Whatever else it says, fine-tuning underscores how mysteriously the two are tangled.

THE COSMIC BODY

For a growing number of physicists, the fine-tuning problem can be solved only by accepting that the entire cosmos is a single, continuous entity, working in seamless harmony like the human body. Everyone accepts that individual cells in the heart, liver, brain, and so on are linked to the activity of the entire body. If you look at a cell in isolation, its relationship to the whole is lost. All you see are chemical reactions swirling in, out, and through the cell. What you cannot see is that these reactions do two things simultaneously: at the local level they keep the individual cell alive, while at the holistic level they keep the entire body alive. One renegade cell that makes a break on its own can become malignant. In the relentless pursuit of its own interests—dividing endlessly and killing other cells and tissues that stand in its way—the malignant cell becomes a cancerous tumor. The breakdown of one cell's loyalty to the whole body is ultimately futile. The cancer meets destruction at the same moment the body dies. Did the universe learn to avoid destruction eons ago? Is fine-tuning a cosmic safeguard that human beings are meant to respect if we hope to survive in the long run?

Let's return to creation stories and myths and look at these questions from their perspective. Myths issue such warnings, beginning long before the chaos threatened by terrorists, hackers, and ecological destruction. In the medieval Grail legends, faith was the invisible glue that held the world together; sin was the cancer that could destroy it. When the Grail knights set out to locate the cup that captured the blood that ran from Christ's side on the Cross, the landscape was gray and dying. Nature's distress reflected human sin. The Grail was a real object, not just a symbol of salvation, and so it was understandable to a population that had almost no learning. In many ways, faith was an invisible link with the Creator. If the Grail could be held up before the people's eyes, that link would prove that God hadn't abandoned them, and the natural order would be upheld.

A single isolated object reverberated through an entire religion—one might say an entire worldview. Another quip from Sir Arthur Eddington applies here: "When the electron vibrates, the universe shakes." Everything in the cosmos is knit together (as perceived by the human brain), because the same reality is at work. If there is another reality "out there" beyond human perception, for all intents and purposes it doesn't exist.

One color-blind person doesn't make colors unreal—there are enough people who can see colors to verify that they exist. But if all people were color-blind, the existence of color would not be perceived by our brains. Humans don't happen to see the wavelengths of infrared and ultraviolet light that lie beyond the ability of our eyes. We can confirm their existence only by using instruments that are designed to detect those wavelengths. When the "darkness" of the universe contains no light or measurable radiation, reality becomes much more like a radio band where we can pick up only one station—the one we recognize as our universe.

Looking back at the early universe, during the phase when atoms began to appear, quantum theory holds that every particle of matter was balanced by a particle of antimatter. Potentially they could have annihilated each other, making the life of the cosmos a very short story. But as it happens—a phrase you're growing used to—there was a tiny fraction more matter than antimatter, calculated at around 1 part per billion. This was precisely enough to allow all the visible matter in creation to escape annihilation, giving rise to the present universe.

A SIDE MYSTERY: FLATNESS

Fine-tuning, when broken down into constants, looks abstract and mathematical. But as with every cosmic riddle, there is visible evidence all around us in physical form. A spectacular example is known as the flatness problem, a side mystery that deepens the main mystery of fine-tuning. Pushing the limits as close to the beginning of creation as possible, great strides were made in the inflationary model discussed in the last chapter. The generally accepted version of this model was devised by theoretical physicist Alan Guth at Cornell, in 1979 (published in 1981). According to Guth's description, the universe began to expand not quite exactly at the instant of the big bang, but a tiny fraction of a second after it.

The evidence for the early universe inflating with remarkable speed comes from various clues. One is the near uniformity of the radiation that emerged during the big bang and continues to pervade the universe today. Another is the near flatness of space. *Flatness* is a technical term in physics that refers to the curvature of the universe and the distribution of matter and energy in it. Newton developed a theory of gravity treating it as a force, which is only one way to look at it. As developed by Einstein, general relativity describes gravity in terms of three-dimensional geometry, so that stronger or weaker gravitational effects can be graphed as curvature in space. The more mass and energy involved, the greater the curvature.

The curve can go both ways; inside, which produces a sphere

like a basketball; or outward, producing a flaring object like a horse saddle. Physics refers to these as positive and negative curvatures. A basketball and a saddle can be modeled as two-dimensional surfaces, but the curvature of space, occurring in three dimensions, is more complex: a ball, for example, has an inside and an outside, while the universe doesn't. General relativity can compute how much mass-energy within a given space causes it to curve one way or the other. If our universe exceeded a critical value, it would have curled up into a ball that shrank to a point and vanished, or, in the opposite direction, would flare outward infinitely. The average concentration of mass-energy has to be very close to this critical value in order to produce the universe we see, where space on large scales is flat.

Because the infant universe was almost infinitely dense, its expansion could only lessen the density, like a lump of taffy that gets thinner as you stretch it. At the current age of the universe, the density of mass-energy per unit of space is quite low, the equivalent of about 6 hydrogen atoms per cubic meter of space. Looking at the overall picture, the present universe appears quite flat. But there's a glitch. The equations of general relativity tell us that if the critical value ever did fluctuate, even by a small amount, the effect in the early universe would be enormously magnified very quickly. Clearly the infant universe held close to the critical value, which is fortunate if you want the universe to exist as it does today rather than being saddle-shaped or collapsing in on itself. But calculations show that the early universe must have had a density extremely close to the critical density, departing from it by not more than one part in 10^{-62}, or one part in the very large number 1 followed by 62 zeros. How was such mind-bending accuracy possible?

Alan Guth's solution, which became accepted as part of the standard model, was to invoke an inflationary field that has a certain density that never changes, unlike the universe that emerged, whose density changes as it expands. (By rough analogy, a lump of taffy can be stretched very thin, but it will always taste sweet. Its taste is "flat" everywhere, no matter what size the taffy is.) In effect, the inflationary field was like a grid that kept the infant universe on a steady course even under the extreme conditions of near chaos. As a result, we see near flatness today everywhere we look. (In a related paper from the same period, Guth gave a field-based solution to another conundrum, known as the horizon problem, which has to do with the even temperature found throughout the universe. We won't go into it here, since the flatness problem illustrates fine-tuning so vividly.)

If physics ever discovers how to integrate quantum theory and gravity, it may one day completely account for the inflationary scenario. The basic tenet is that wrinkles of space in the quantum field (or vacuum) eventually formed the visible universe and its array of galaxies. These wrinkles or ripples could have been produced by extreme gravitational forces microseconds after the big bang—see pages 16–17 for our earlier discussion. What happened before the inflation is less certain; to account for the Planck era requires theoretical developments that for now are out of reach.

WHAT IF FINE-TUNING HAS TO EXIST?

The reliance on randomness in so many current theories feels intuitively suspect when we consider the beauty and complexity of creation. Why was physics persuaded to go down this road? Despite the fact that *design* is an abhorrent word to cosmologists, it's very hard to look at fine-tuning without suspecting hidden patterns, and once this happens, you're forced to ask where these patterns came from, if everything is supposedly random.

In the last century, Eddington and Paul Dirac, a physicist, first noticed that certain coincidences in dimensionless ratios can be found. That is, instead of applying only to very large dimensions or very small ones, these ratios link microscopic with macroscopic quantities. For example, the ratio of the electric force to gravitational force (presumably a constant), is a large number (Electric Force/Gravitational Force = E/G ~ 10^{40}), while the ratio of the observable size of the universe (which is presumably changing) to the size of an elementary particle is also a large number, surprisingly close to the first number (Size Universe/Elementary Particle = U/EP ~ 10^{40}). It is hard to imagine that two very large and unrelated numbers would turn out to be so close to each other.

Dirac argued that these fundamental numbers must be related. The essential problem is that the size of the universe is changing as the cosmos expands, while the first relationship is presumably unchanging, since it involves two supposed constants.

To make this less abstract, imagine that you were born three miles from your best friend. All your life you remain best friends—a constant—and whenever you move to a new house, so does your friend, and always the two houses are three miles apart. Moving from house to house is the changing part. In the human world your friend can decide (for some strange reason) that the distance between you must be three miles. But how does nature "decide" to match the relationships Dirac discovered? Dirac's *large number hypothesis* was a mathematical attempt to link ratios in such a way that they aren't coincidental.

But wasn't the anthropic principle accomplishing the same thing? It didn't use advanced mathematics but instead a chain of logic that can be intuitively grasped. A Martian landing at Yankee Stadium wouldn't have much chance of grasping the rules of baseball just by watching the game but could infer that all the players had a connection—the rules of the game—that guided every move. If you don't know the rules, watching a batter bunt instead of taking a full swing looks random, as do many other actions, like whether a runner tries to steal a base or not. The anthropic principle tries to make a similar point. Even if we, like a visiting Martian, can't figure out the rules by examining the universe directly, its precise movements tell us that some connection must be guiding the game.

The anthropic principle holds a special fascination for the two authors of the book you are holding, because it is a step toward the possibility of a human universe. However, there is a troubling flaw that dampens our enthusiasm, namely, coincidences aren't science. Even the most remote coincidence isn't science. On rare occasions, for example, two people meet on the street or at a party who look almost identical. Or a person may look so similar to Elvis Presley that he works as an Elvis impersonator. The coincidence is striking, but it is false logic to claim that there must be a deeper reason for why it exists.

If you think about it, the anthropic principle just states the obvious: "We are here because the conditions were right for getting us here." There is no explanatory power in that sentence. It's a little like saying, "Airplanes fly because they can lift off the ground." Even so, nothing in current physics offers an explanation that overrides the anthropic principle.

One way to deal with the flaws of the anthropic principle is to counter that the constants have shifted as the universe evolved—and are still shifting. But that's a queasy possibility. It's more comforting to believe in timeless constants, which will never rock the boat. You can take the gravitational constant and the speed of light (c) in the formula $E = mc^2$ to the bank.

But their stability could be an illusion, and *illusion* isn't a comforting word. If you get rid of fixed constants, how does one live? How does one get to work or fight an infection with antibiotics or balance a checkbook unless we embrace illusions? The answer is that we live better. Timeless constants don't have to be chucked out the window; we only need to see through them, realizing that in a participatory universe, human beings have a higher status than numbers, however advanced the mathematics. In a human universe, the constants shift to accommodate us, not the other way around. That's a huge claim, we know. Right now, we're building a case for it, and the present order of business is to show that even the best answers in current physics have insurmountable problems unless we change our worldview.

PICKING A PATH FORWARD

As far as this book is concerned, the fine-tuning problem boils down to two clear choices. On the one hand, fine-tuning is a case of coincidence piled higher and higher, and the only explanation is that humans just happen to exist in the right universe by chance. This is the viewpoint favored by multiverse and M-theory proponents, including Stephen Hawking and Max Tegmark. They accept the possibility of nearly infinite universes churning out every possible combination of constants, zillions of which do not match in such a way as to form life. But one did, and we happen to live in it. This is the cosmic equivalent of putting a hundred monkeys to work tapping randomly on typewriters, eventually producing the complete works of Shakespeare (after also producing an almost infinite mountain of gibberish). Pure randomness rules if we live in nothing more than an incredibly unlikely universe of our own—lucky us.

How lucky are we, exactly? Estimates that dovetail with superstrings (assuming they even exist) yield 1 chance out of 10^{500}—that's one part in the extremely large number of 1 followed by 500 zeros—10^{500} being a number far greater than the number of particles in the known universe. A hundred monkeys writing all of Shakespeare is a million times more likely; they could even write the rest of Western literature while they were at it. But it gets even more cumbersome. From so-called chaotic inflation theory, the chances of being in the right universe are much smaller, $1/(10^{10})^{10})^{7}$! It's one thing to claim that a hundred monkeys can write Shakespeare if given enough time; quite another to declare that there is no other way for Shakespeare to be written, which is the claim M-theory and the multiverse hypothesis are making. (Actually, the central claim of the multiverse is far more radical, since it states that all the possible laws of nature unfold in infinite ways, infinite times over. Probabilities break down when the odds for and against anything are both infinite. As Alan Guth puts it, here on Earth two-headed cows are rarely born, but we can compute the odds of that happening by assigning a number to specific mutations. In the multiverse, however, one-headed cows and two-headed cows are both infinite in number, so computing anything about them falls apart.)

We said that two clear choices exist. The other choice, which we favor, is that the universe is self-organizing, driven by its own working processes. In a self-organizing system, each new layer of creation must regulate the prior layer. So the generation of every new layer in the universe, from particle to star to galaxy to black hole, cannot be considered random, given that it was created from a preexisting layer that in turn was regulating the layer that produced it. The same holds true throughout nature, including the workings of the human body. Cells form tissues, which in turn form organs, the organs form systems, and finally the entire body has been created. Each layer emerges from the same DNA, but they stack up, as it were, until the pinnacle of achievement, the human brain, crowns it all.

Yet, as magnificent as the brain is, compared with a single colon cell, the smallest component in its multilayered structure is cared for and nourished. DNA has evolved this skill at building hierarchies because the entire universe was its schoolroom. This *recursive* system of self-organization, to give it a scientific name, where every layer curves back on itself to monitor another layer, pervades physics and biology.

For example, your genes produce proteins that monitor and regulate the entire genome, tending to repairs and mutations in your DNA. In your brain neural networks create new synapses (the connecting gaps between brain cells) that in turn monitor and regulate the preexisting synapses that gave rise to them. The brain integrates all new knowledge, information, and sensory input by associating it with what you already know. Whether we're speaking of genes and the brain or solar systems and galaxies, self-organization is present. Existence requires balance, which demands feedback. By monitoring itself, a system can correct imbalances automatically. Every new bit of the universe, however minuscule, must create a feedback loop with what gave rise to it. Otherwise it wouldn't be connected to the whole—in human terms, it would be homeless.

Viewed this way, fine-tuning isn't a mystery. No one finds anything mysterious about how the gear train in a car's transmission fits together precisely. If it didn't, the car couldn't operate. In the same way, an operational universe must be fine-tuned. Why would we expect the opposite to be true, that the universe is naturally ramshackle? What's actually natural in nature at every level is self-organization. Even when an event appears to be random (satisfying the mathematics of randomness), a kind of purpose is invoked, beginning with the overarching purpose of homeostasis, the dynamic balance of all parts in a whole.

In high-school biology, the classic example of homeostasis is the body's ability to maintain a steady temperature of 98.6 degrees Fahrenheit under changing thermal conditions outside. Let's say that you've been caught outdoors in the fall without a jacket as the temperature suddenly drops. Depending on how long you are exposed, your body will go through a series of tactical steps to ensure that your vital organs don't get chilled, such as moving blood away from the skin closer to the bodily core and revving up your metabolic furnace. Under a microscope, the activity of any single cell might look arbitrary and random—until you realize what the entire body is trying to accomplish.

In our view, the fine-tuning of the universe shows how sensitive nature is, balancing galaxies by making sure that subatomic particles are in balance first.

Self-organization is embedded in the fabric of the cosmos, *acting like* an invisible, offstage choreographer to drive evolution—but this mustn't be mistaken for the red herring of "intelligent design" by a supernatural God in the sky. The smooth running of the universe is underpinned by quantum processes, rapidly making invisible, microscopic choices that lead to final results at the level of daily life.

Do humans exist on our planet as winners in a cosmic game of roulette, overcoming incredibly small odds of finding the right universe? Or do we exist because we fit into the hidden scheme of nature? Most people answer according to their worldview, which can be religious, scientific, or a blurry hybrid of the two. Yet one thing is certain. If we believe in a hidden scheme or a grand design, we'll see it "out there."

We participate in the universe by finding order and figuring out where the patterns come from. Einstein hit upon a deep truth when he said, "I want to know the mind of God; everything else is just details." Substitute "the purpose of the universe" for "the mind of God" and you have a goal worth pursuing for a lifetime.

WHERE DID TIME COME FROM?

Time was never meant to be our enemy. We have turned it into one, by saying things like "I'm running out of time" or "Time's up," which implies that human beings are trapped in the prison of time with no chance of escape, at least not until death reveals if the hope of an afterlife is true. Einstein found a way to make peace with time, however, when he said that past and future are illusions; only the present moment exists. This is one of those brilliant moments when the world's spiritual traditions and advanced science converged. Did an enlightened sage, a prophetic poet, or a famous physicist say the following: "For eternally and always there is only now, one and the same now; the present is the only thing that has no end"?

The words are those of Erwin Schrödinger, who, like many quantum pioneers, drew closer to mysticism the more he understood the revolution he had helped to create. Since the "mystical" has fatal effects in science, what happens if we decide that Schrödinger was being completely literal? We would be left with a now-familiar mismatch. Time in everyday life definitely moves from past to present to future. How can it be that time stands still, or, even more incredible, that time was invented by the human mind?

Go back in your mind to the childhood image you had of

Heaven. Whether you see angels playing their harps on clouds or green pastures with innocent lambs gamboling through them, every child is told that Heaven is eternal—it lasts forever. To a child's mind—and the minds of many adults—eternity sounds boring and monotonous. Ultimately it might even be frightening, as time endlessly unfolds and harp playing and lambs lose their appeal.

But eternity doesn't in fact last a long, long time. Eternity is timeless, and when any religious faith promises eternal life, two things are involved. One is the absence of time's afflictions, such as growing old and dying. The second promise is much more abstract. After death we become timeless. Literally without time in the "zone of eternity" where souls abide. But why wait for an afterlife? If time is an illusion, we should be able to step out of it whenever we want, simply by living in the present moment—then the value of going to Heaven will be achieved.

Scientists don't think this way—most of them, at least—but it was science that opened the door to seeing time in a new way. No one knew, for example, that time could stretch like a rubber band until Einstein pointed it out. Spiritual teachers had already told us that God's time is infinite, and now some cosmologists are saying the same thing about the multiverse. In fact, modern physics is very greedy to capture more and more time. If there is literally infinite time, then infinite universes could spring up, and if you have infinite universes, there could be a mirror image of Earth "out there" somewhere, with mirror images of all the people alive today.

All of these speculations, including the religious ones, are fanciful until we know where time came from. There is no proof that the big bang *took any time at all.* That's because when you dive into the pure chaos of the Planck era, time was just another ingredient in the quantum soup, swirling around with no properties like "before" and "after" or cause and effect. The universe was once a timeless place—perhaps it still is.

GRASPING THE MYSTERY

The most accurate atomic clocks are so precise that every few years a "leap second" must be thrown in—the newspapers insert a little story when this happens, the last occasion being June 30, 2015. The need to add an extra second arises because Earth's rotation is gradually slowing down, and adding the extra second brings Coordinated Universal Time (clock time) back in sync with solar time (sunrise and sunset).

When clocks based on the vibration of atoms can slice time into millionths of a second, it would appear that time doesn't have many mysteries left. Clocks are very useful for telling time. But they also conspire to keep us from knowing the truth about time. When asked to explain relativity, Einstein famously said, "Put your hand on a hot stove for a minute, and it seems like an hour. Sit with a pretty girl for an hour, and it seems like a minute. That's relativity." He was slyly referring to the personal aspect of time, and that is where the hidden mysteries begin. When someone is feeling blissfully contented, they often sigh, "I wish this moment could last forever." Are they wishing for something that could already be real?

Because time has two faces, one relating to personal experience and one relating to the objective world described by scientific equations, the issue is a tangled one. No matter how time seems to drag in the dentist's chair or in a traffic jam, the time registered by a clock isn't affected. You can slice this fact two ways. You can claim that clock time is real, while personal time isn't. Or you can point out that subtracting the personal aspect of time is possible only in theory. In the world of experience, all time is personal. We take the second position, even though it sounds radical and even peculiar at this point.

When time gets intensely personal, we notice the human element that typically hides out of sight, because we take it for granted. Shakespeare's Macbeth is at his most despondent, having killed a king and setting his own tragic fate in motion, when he wearily declares, "Tomorrow, and tomorrow, and tomorrow, / Creeps in this petty pace from day to day, / To the last syllable of recorded time."

This is a classic expression about the personal aspect of time. One day inexorably follows after another, bringing us closer and closer to the moment of death. But time's "petty pace" is actually an illusion. Time doesn't "flow" in the quantum field, where all of reality exists as pure potential. The quantum field is outside our commonsense notion of time, and when a particle emerges from the field, it has no history. Particles are tied to an on/off switch, not to the past.

In a quantum reality, Macbeth would say, "Now and now and now. Nothing else exists but the present." If the flow of time is no longer credible, the only time that can possibly exist is the present moment. The present moment is the measure of "real" time, while the "flow" of time, which produces the birth of babies and the death of old people, is an illusion. There's the rub. We see babies being born and old people dying, among many other things that happen in the flow of time. No one can tell us that these things are illusory.

Naturally, this illusion is very convincing if you happen to be alive on Earth. But to a physicist, the timeless quantum field is being filtered through a human nervous system, which cuts eternity into neat, practical slices for our own benefit. "Out there," time is a dimension of reality totally detached from human concerns. Macbeth may be afraid to die, but a magnet isn't. It exists in the electromagnetic field, which for all practical purposes never ages. For as long as the present universe endures, the electromagnetic field remains intact, never growing old. A lightbulb burns out after a certain number of hours, but light itself doesn't burn out. Even if the cosmos should reach an endpoint billions of years from now, and every source of light goes dark, it would be wrong to say that light got old. It would simply shut off.

COSMIC CHICKEN OR COSMIC EGG?

To a working scientist, this position seems so self-evident that you'd think it couldn't be challenged. But almost immediately we run into a "which came first, the chicken or the egg?" dilemma. You can't have time without the universe, and you can't have the universe without time. The two depend upon each other. The same is true for the atoms, which didn't appear until 300,000 years after the big bang, when bare protons and electrons combined; before then, only ionized matter existed. Without time, there would be no atoms. But without atoms, there would be no human brain to perceive time. How did the two get linked? No one knows. The illusion created by clocks can't be trusted, which casts doubt on objective time itself. Something insurmountable, a Chinese wall, prevents us from peering beyond the Planck era into the pre-created state and seeing what came before the big bang. The same wall exists with respect to time, but this hasn't stopped physicists from reaching for an explanation of how time operates in the created universe. Time brings change, and change implies motion, which can be observed everywhere in creation. But strangely enough, motion doesn't mean that we are observing something moving. This, too, could be an illusion.

The fact that atoms and molecules move around is part of the clock illusion. When you watch a car chase in the movies, the cars aren't actually moving. Instead, frames of still photos spool through the projector (when projectors used film) at twenty-four frames a second to create the illusion of motion. Our brains also operate by taking snapshots—fixed images—and stringing them together so quickly that we see the world in motion.

At the level of the quantum field, all motion is deceptive. Subatomic particles wink in and out of the quantum vacuum, reappearing each time in a slightly different place. Essentially they don't move, because the different places are just changes of state. Think of how a TV screen works. If a red balloon needs to pass across the screen, nothing inside the TV has to move. Instead, the phosphors (in an old-fashioned cathode ray tube) or the LCD lights (in a digital screen) wink on and off. By doing that in sequence (first red LCD number one, then red LCD number two, red LCD number three, and so on), the balloon appears to float from left to right, from up to down, or any way you choose.

Sitting at the movies, we may know how the trick is done, but we give in to the illusion. Any time we want, we can get up and walk out of the cinema, returning to the real world. But how do you walk out of the real world? If everyday time is just as illusory as movie time, there's a problem. The human nervous system is constructed of tiny clocks that regulate other tiny clocks all around the body. Besides the really big rhythms the body follows (sleeping and waking, eating, digesting, and excreting wastes), there are medium rhythms (breathing), short rhythms (heartbeat), and very short rhythms (chemical reactions inside our cells).

It's a miracle that the human nervous system can synchronize all of these rhythms, and more. There are the twitches of muscle fibers, the flow of hormones, the division of DNA, the production of new cells—all these processes have their own clocks. DNA activity also controls long-range rhythms, from the emergence of baby teeth, the start of menstruation, and other aspects of puberty to more distant events like male balding, menopause, and the onset of chronic illnesses that take years to develop, such as many cancers and Alzheimer's disease. How our genes manage to span timescales as short as a millisecond (the time a chemical reaction inside a cell might take) and as long as seventy years or more remains a mystery.

At this point, if you are practical-minded, you might be

tempted to say, "The mystery of time is too abstract. As long as my brain is running things by the clock, that's good enough." But it isn't. Imagine that you are in bed, dreaming. In your dream you're a soldier fighting on the battlefield. You charge across the field, your heart pounding. All around you, bombs are going off; artillery shells whiz overhead. The spectacle rivets you even in your terror—and then you wake up. At that instant, everything in your dream is revealed to be an illusion, but most especially time. In our dreams, long spans of time can pass, but neurologists know that the episodes of REM (rapid eye movement) sleep, where almost all dreams occur, take no more than a few seconds or minutes.

In other words, there is no relation between "brain time" as measured by neural activity and the experiences inside a dream. The same is true when you're awake, however. See yourself sitting by a window in a dream watching people and cars go by. When you wake up, a dream researcher tells you that your dream, which felt like half a day, in fact took twenty-three seconds of brain time. If you sit by a window watching the passing world while you are awake, that experience is also created by the same brain cells that create dreams. The firing of a few neurons, which takes only a few hundredths of a second, can cause you to see a bright flash in your eyes that lasts a long time (seeing such lights is common in conditions like migraine and epilepsy). You have a choice whether to call brain time the real thing or your experience the real thing. But in actuality neither is more real than the other, for the simple reason that we can't step outside our brains in order to capture real time. Walking out of a movie is easy. Walking out of this waking dream isn't.

So, how does the brain learn to keep time? We could look to the chemical reactions taking place inside brain cells, which like all other cells are chemical factories. These reactions, along with the electrical activity that "lights up" on an fMRI scan, are precisely timed. One crucial activity is the exchange of sodium and potassium ions across the outer membrane of a brain cell. (An ion is an electrically charged atom or molecule, either positive or negative.) The time this takes is infinitesimally small, but it's not instantaneous. So there's your basic brain clock, or a key part of it.

Unfortunately, the brain's clock isn't attached to the experience of time. While all those ions are clicking away, time can be behaving any way it wants in dreams, hallucinations, under disease conditions, in moments of inspiration, or other uncanny moments when time stands still. Clicking ions tell us nothing about the behavior of time, and anyway, there would be no ions in the first place without the big bang. We are at the same dead end where the mystery began. The cosmic chicken-and-egg question is still up for grabs.

OR MAYBE NOT . . .

The so-called dead end has actually revealed an important clue. Time is springing into existence with every firing of a neuron in the brain. Its creation is constant. For as long as a person is alive, he or she is "creating" time; we never run out. (When someone says, "I ran out of time," of course they really mean that they didn't meet a deadline.) Therefore, we don't have to go back to the big bang. To ask where time came from isn't really about the universe. It's about our experience here and now. *There is no other time.* Solving the mystery of time will tell us if humans are the creators of time or its unwitting victims, the pawns of brain activity. There seems to be no other choice. If time depends on the brain and vice versa, we are talking about one of the most important ways that every person participates in the universe. Before relativity, the belief that everyone shares the same experience of time formed a kind of cosmic democracy. We were all equal in how time operated. This condition can be called a

Galilean democracy (after the great Italian Renaissance scientist Galileo Galilei), because of some crucial observations by Galileo that reinforced commonsense reality. For example, if someone going past you in a car throws a ball moving in the same direction, the speed of the ball can be reliably calculated, and the result will always be the same. A car moving at 60 miles per hour might have a Major League pitcher as a passenger. If he throws a fast ball at the record speed of 105.1 mph (set in 2010 by Aroldis Chapman of the Cincinnati Reds), the actual speed of the ball will be 165.1 mph, which is arrived at by adding the speed of the car to the speed of the ball.

The Galilean democracy was good enough as long as there was a fixed point to stand on. To the pitcher in the car, the ball only travels 105.1 mph, because he is already moving as fast as the car is. However, Einstein pointed out that there is actually no fixed spot in the universe for measuring time. Every observer is in motion relative to every other observer. (No one can prove for sure who is moving and who is not moving, at least for constant motions.) Therefore, all measurements are relative, depending on how fast two things are moving past each other.

Relativity toppled the Galilean democracy. A reality equal for all participants in it was no longer a reliable possibility. If you are in a spaceship traveling at the speed of light, and you shoot a ray gun off your port bow, the photons from your gun would also travel at the speed of light. Unlike the baseball pitcher in a speeding car, you can't add the speed of your spaceship to the speed of the photons you're firing. By traveling at the speed of light, you are already at the absolute limit for all observers in all moving frames of reference. Einstein showed that the rate of time passing would depend on the frame of reference one is in. Thus relativity dismantled forever the assumption that everyone's experience of time is the same. Time is not universally the same for every observer. We are like free-floating points in space where only local time applies.

But if you look at it another way, every observer defines the time frame he is experiencing and can change that time frame by moving faster or slower, in a sharper curve, or by approaching a strong gravitational field. The Galilean democracy has turned into an Einsteinian democracy.

In fact, it's a universal democracy, which has brought with it more freedom of participation. The constants are still there. The speed of light will impose the same limitation on how fast an object can move through space-time. But instead of acting like a prison wall hemming us in, the constants are like the rules of a game. You must play by the rules, but as long as you do, you can make any move you want, whether the game is chess, football, or mah-jongg. Science has a tendency to lean too far toward the rules. Since electromagnetic waves travel at the speed of light in empty space, for example, they won't change speed anywhere in the cosmos. Fixing the speed of light as an absolute was a desirable achievement from the standpoint of making calculations, because it removed the unreliability of subjective time.

The scientific viewpoint, which says that the brain is bound by the speed of electric currents, is just that, a viewpoint. In Einstein's democracy, each person is free to put the rules first or freedom first. There is no absolute place to stand. The constant speed of electromagnetic waves is a boundary our brains must respect, but our minds are allowed freedom of thought; we can play any mental game we choose—and in the end, all games are mental. The speed of light doesn't constrict our humanity, only our neurons.

When relativity toppled absolute time, it also toppled space. As with time, space looks distorted when measured at different moving frames of reference. According to relativity, a stationary observer watching a spaceship approaching the speed of light would see its length being shortened in the direction of its forward motion. In everyday life we don't subjectively perceive these relativistic effects in space and time, because the speeds we're accustomed to are tiny compared with the speed of light. In particle accelerators, however, such as the Large Hadron Collider (LHC) in Geneva, Switzerland, where the Higgs boson was discovered, it's routine to accelerate subatomic particles to speeds approaching the speed of light. That's one place on Earth where relativistic effects are measurable and totally accepted as a fact of nature by the experimenters.

In a word, we can visualize what time could be like as it enters creation. Think of pop-up books, which lie flat like ordinary books but when opened, suddenly open up into a house, animals, an elaborate landscape, and even have moving parts. Creation is like that, when viewed at the quantum level. There's flatness, and suddenly there are objects in space-time. Everything pops up at once. Therefore, the isolated behavior of particles isn't really indicative of reality. In order for a tree, a cloud, a planet, or the human body to exist, there isn't really a piling up of subatomic particles, atoms, and molecules the way bricks are assembled to build a house. Instead, the subatomic particles *bring space and time with them.*

This fact has astonishing implications. For example, a particle moving close to the speed of light may decay in a short amount of time, measured in millionths of a second, but it will last longer as observed by physicists in a laboratory that's stationary with respect to the moving particle. A particle that moves exactly at the speed of light lasts forever, because for it time doesn't pass. It seems to stand still. As far as light is concerned, time doesn't exist, while from our perspective, in a world cordoned off by the speed of light, the lifetime of a photon is infinitely long. Photons, the particles of light, have zero mass. If a particle (any particle) has a finite mass, it can never reach the speed of light.

Now we have proof for one of the seemingly impossible ideas this chapter started with: eternity is at our doorstep. Light, which is timeless, gave rise to life on Earth and continues to sustain it. Therefore, the real question is how two opposites, time and the timeless, relate to each other. Time, space, and matter spring out of flatness all at once, and when solid objects get dragged into the Einstein democracy, they become relative. According to relativity, the mass of an object isn't constant. Matter is constantly being transformed into energy and vice versa, as $E = mc^2$ verified. But here our ability to visualize breaks down. We are limited by the slowness of the brain, just because it is made out of matter. The electrical impulses inside the brain travel at great speed, but the thoughts they trigger are "stepped down," like the enormous voltage in power lines being stepped down for household use. The only particles that move at exactly the speed of light are photons and other particles that have zero mass, such as the elusive neutrino, if indeed it has zero mass. If one could magically exceed the speed of light, time would run backward, a theoretical doorway back to the beginning of time.

Einstein reasoned that this couldn't happen in a classical world, even one with relativistic effects. However, it may happen in a quantum world. All the permutations of time are quantum possibilities, which offers another valuable clue. If the quantum domain allows for time to stand still, move backward, or follow the arrow that moves from past to present to future, then there is no reason why the big bang favored any of these possibilities over the others. Asking why we happen to live in clock time is very much like asking why the universe fits together so perfectly. Clock time benefits human beings, just as the fine-tuned universe does.

As with all life forms, humans cannot exist without birth and death, creation and destruction, ripeness and decay. These are the gifts of clock time, and although stars and galaxies also undergo birth and death, their life cycles are only a matter of shuffling matter and energy around on the cosmic game board. The human situation is much more complex, because unlike physical objects, we have mind, which creates new ideas born in a field of possibilities that seems infinite. The mystery of time somehow must be linked to how the human mind works. Let's see if the quantum revolution brought time and mind closer together.

ARE QUANTA ON THE CLOCK?

Going faster than the speed of light would be very embarrassing for relativity, and now it has happened. Recently experimenters have found a way to move photons from one position to another without passing through the space in between, the first example of true teleportation. Because the photons skip from point A to point B instantaneously, no time elapses. The speed of light isn't actually exceeded; it's made irrelevant. We may say that time is bypassed. In fact, teleportation unravels the neat pop-up picture of space, time, and matter.

Teleporting photons have enormous implications. Einstein's thinking, as we've been discovering, remained rooted in a classical world, and such a world is bound by the speed of light. Like wild horses released from a corral, if quantum objects can go beyond the speed of light—not by traveling faster but through instantaneous action—something unknown lies ahead.

One unknown has to do with how many dimensions actually exist. Clock time is one-dimensional. It travels in a straight line that occupies one dimension, as all straight lines do—they can only connect point A and point B. But in quantum theory there is no limit to how many dimensions there are, since they exist as purely mathematical constructs. For example, a number of quantum theories require us to go beyond gravity to the field of supergravity, which posits eleven dimensions. The pre-created state before the big bang could be dimensionless (occupying zero dimensions, mathematically speaking), or it could have infinite dimensions. The possibilities are head-spinning, being so far removed from everyday experience.

We have to add our universe's three dimensions to the pile of dismantled absolutes, and time, the fourth dimension, may have to go with it. In mathematical terms it already has. It is generally accepted that every particle is emerging here and now from a place of zero dimensions, namely the quantum vacuum. Some radical physicists even theorize that the only two numbers with any reality are zero and infinity. Zero is where the trick of turning nothing into something occurs. Infinity is how many possibilities can emerge on an absolute scale. Every number in between has the reality only of soap bubbles and smoke.

Zero dimensions cannot be visualized—and even the mathematics can seem like a parlor trick, because so many variables are either unknown or sheer guesswork—but surely all of us exist because the timeless, which has no beginning or end, expresses itself as time in the present moment. This transformation defies logic, which should come as no surprise by now.

Since the quantum realm isn't on the clock, why not accept the truth—that time is totally malleable? In that case, it's no great leap to seeing any version of time itself as artificial. To make this easier to comprehend, we need to explore a basic term in quantum physics that also applies to everyday reality: *state*. When you see a tree, its state is that of a tangible object you can locate in space-time and experience with your five senses. A floating cloud is vaporous and more elusive than a tree, but it exists in the same state of physicality.

When physics delves into the quantum domain, another state is involved, the virtual state. It is invisible and intangible but nevertheless real. In fact, we visit the virtual state every waking moment. Think of a word, any word. We'll pick *avocado*. When you think or say "avocado," it exists as a mental object. Before you think or say the word, where is it? Words aren't stored in a physical state in brain cells; instead, they exist invisibly but ready at hand—in a virtual state. You can pluck them out at will, an ability that deteriorates when the brain's memory retrieval gets physically weakened or damaged. A faulty radio can't retrieve radio waves, either; without a working receiver, radio signals exist, invisible and unsensed, all around us.

Likewise, the brain is a receiving device for the words we use, and not only that; the rules for using language are also in the virtual domain. When you see the sentence "Are house need wind?" you instantly know that it doesn't obey the rules of language. You use no energy inside your brain to tell the difference between sense and nonsense. The rules are invisibly embedded in a place that is, for all intents and purposes, nonphysical. Subatomic particles also come from a place that is nonphysical, and there's no reason to believe that where you go to fetch the word *rose* isn't the same place from which galaxies spring.

The virtual state lies outside the manifest creation. When a wave turns into a particle, which is the basic step that brings photons, electrons, and other particles into the world of our experience, the virtual state is left behind. The virtual state is also why physics computes that every cubic centimeter of empty space isn't actually empty. At the quantum level, it contains a huge amount of virtual energy.

All things in the universe can change states. In everyday experience, no one is mystified to see water turn into ice or steam, which are other states of H_2O. At the quantum level, changes of state reach their limits, poised between existence and nonexistence. A kitchen table is transiting from the virtual to the manifest state thousands of times a second, too fast for anyone to observe. This is the winking in and out, or on/off switch, that we've mentioned several times. A quantum change of state is *the* basic act of creation. The multiverse gained enormous popularity for this very reason, when it was realized that a universe popping into existence was no more eventful than an electron popping into existence. The same fluctuations in the quantum field were at work. To the naked eye the universe looks very, very big, while an electron is very, very small, but this difference doesn't matter in the act of creation.

A quantum popping into existence doesn't come from somewhere "else," nor does it go anywhere. It is only a change in state. Therefore, instead of using time as the measure of change, we must think in terms of states. Think of a volleyball tethered to a pole. When you hit it, the ball starts revolving around the pole, but at a certain point it runs out of energy and comes closer and closer to the pole, finally reaching a state of rest. (Planets orbiting the sun would fall into it if they lost energy and momentum over time, except for the fact that they travel in the vacuum of outer space. Unlike a volleyball, they meet no air resistance and thus can keep spinning round for eons.)

Now imagine an electron orbiting around the nucleus of an atom, an image that appears very similar to a volleyball circling a pole. With atoms, each electron orbit is called a shell, and electrons stay inside their assigned shell unless a quantum event occurs, in which case they jump a shell closer or a shell farther out. The word *quantum* was assigned because, as a "packet" of energy, the quantum travels from one definite state to another, carrying its energy with it. Electrons don't slide from one location to another, nor do they slow down. They pop out of one orbit (shell) and appear in another.

When you grasp the importance of "state," you see why quanta aren't on clock time. Clock time is like ticker tape spooling out of the tape machine continuously, while the quantum domain is full of gaps, sudden changes of state, simultaneous events, reversals of cause and effect. So, if the basis of creation is quantum, how did physical objects get tied to clock time in the first place? The simplest answer is to say that clock time is merely another state. Once the universe matured, around a billion years after the big bang, every gross physical object (i.e., bigger than an atom) was locked into the same state of manifestation. Advanced mathematics, using probability theory, can compute the very, very remote chance that a kitchen table might totally disappear into the virtual domain, only to reappear three feet away.

But that's not a practical consideration. Being locked into manifestation, gross objects in the everyday world are reliable in their subjugation to space-time. Despite the quantum's vanishing act as it winks in and out of existence, kitchen tables aren't going anywhere soon on their own.

So the real question is, how do changes of state occur? The big bang, which caused the entire universe to emerge instantaneously, was a change of state that can't be explained as happening in one place or at one specific time. During the Planck era, everywhere and nowhere were the same, as were before and after. Despite the wall that prevents us from witnessing the Planck era, we could call it a phase transition whereby one state transformed into another and the virtual became manifest. It's quite peculiar, sitting here where the clocks tick, to realize that, just like an electron popping into a new shell, the entire creation did the same almost 14 billion years ago. But if we can imagine it, at least this tells us how something as tiny as an electron and something as large as the cosmos are linked. Neither one is on clock time; therefore, entirely new ways of thinking must be adopted.

PSYCHOLOGY MAKES AN ENTRANCE

Now we're ready to take you, personally, out of the prison of time. Your body participates in the universe through changes of state. Let's say a stranger knocks on your door one day. You open it, and he introduces himself. If he says, "I'm your long-lost brother, and I've spent years trying to find you," you will go into a different state than if you hear him say, "I'm from the IRS and we are confiscating your house." Your body will react instantly and dramatically in both cases. Simply by hearing a few words, your heart rate, respiration, blood pressure, and the brain's chemical balance turn on a dime.

In human life, a change of state is holistic; like an electron, you can jump to a new level of excitation. A stranger introducing himself might turn your life upside down. Even as you undergo a dramatic change of state, however, you can't observe the microscopic physical processes taking place in your cells. The particular areas of the brain that create joy or anxiety will light up on a brain scan, but subjectively we experience only the final outcome, not the mechanics of getting there.

But one thing stands out: the triggering event—a stranger knocking on your door—is what begins the change of state. It's not the case that the quantum, although often called the basic building block of nature, is actually building the experience. The chain of command, as it were, moves from top to bottom. First comes the stranger at the door, then the words he speaks, your mental reaction, and all the physical stuff. In short, mind comes before matter. Only in the human world are we certain that this is true, despite the grumbling of materialists, who believe that every event, including mental ones, are caused by bits of matter exchanging bits of energy. Words are mental events first and foremost, because their purpose is the exchange of meaning, not the exchange of physical energy. If someone utters the phrase "I love you," the physical stuff of the body reacts one way; if instead one hears "I want a divorce," the physical stuff reacts another way.

This fact wasn't lost on some quantum physicists, including John von Neumann, a brilliant theorist who took the bold step of declaring that the quantum domain, and reality itself, has a psychological component. Nature is dual, both subjective and objective. That's why we humans can see any situation from either perspective. Meeting a stranger at the door, you can measure his height, weight, hair color, and so on (objective) or listen to what he has to say (subjective). Eyewitness reports of crime are notoriously unreliable in court because all of us mix up our viewpoints. Someone who threatens us grows larger in our minds, making it hard to give an objective account of how tall he is.

Von Neumann took the dual nature of reality quite far, to the very essence of how nature operates. He described a reality where quantum particles make choices and where the observer changes the thing he observes. Quantum physics has been swamped with subjective effects for over a century, largely thanks to the uncertainty principle, which holds that the properties of a quantum can't all be known. The observer selects a property, and suddenly that's what the quantum displays. At the same time, its other properties slip away and are even changed simply by being observed.

Though this sounds abstract, here's an everyday example. You are standing on the north shore of Oahu in Hawaii, a place famous for its massive waves and a mecca for high-risk surfing. As a wave rolls in, you take a snapshot to show your friends. The snapshot stops the motion of the wave, which means you can see how big it is but not how fast it was moving. You've selected one property only. When a physicist observes a subatomic particle, he's taking a sort of snapshot that reveals something he wants to measure, while excluding the other properties. It's not satisfactory to look at reality this way, however, since reality is all-embracing. To compensate for the properties that vanish into thin air when a single property is observed, the other properties of a subatomic particle are calculated as probabilities.

In our everyday example, as you are showing around your snapshot of a huge wave in Oahu, someone might ask, "How fast was it moving?" You vaguely reply, "Real fast." If asked to pin down your answer, you know that the wave was moving faster than a snail but slower than a jet plane. Its actual speed is probably between 20 and 60 miles per hour. Since the wave has long since disappeared, all you can work with is this probability. Quantum physics finds itself in much the same position, leaving open a basic question. How much does an observer change the "real" facts?

Von Neumann didn't conjecture on this point. His breakthrough was that reality has a psychological component (the mind-like behavior of subatomic particles) that is essential. Some physicists, such as Schrödinger, have held that the psychological component is paramount. Schrödinger declared as an "absolute essential" that we "surrender the notion of the real external world, alien as it seems to everyday thinking." But materialism, which traces all phenomena to the existence of the external world, hasn't budged. Either the psychological component is denied altogether or it is extracted from the equation.

How does the psychological side of reality affect time? It's well known that traumatic experiences cause time to slow down. Subjects report that in the midst of battle or during a car crash, everything moved in slow motion. In sports, the concept of "being in the zone" is an altered state where the player can do no wrong, where everything meshes perfectly, and in addition, the world grows silent and time slows down. Athletes report being in a kind of dream state divorced from everyday reality.

It's hard to see how these reports can be sorted out to remove the subjective component. However, experiments have been successfully done in a more controlled environment. In one study, subjects took an amusement park ride that dropped them from a high tower. They experienced free fall before a parachute opened and gently lowered them to the ground. When asked how long they were in free fall, subjects always exaggerated the time, the same way people do in any traumatic situation. The actual time they were falling can be measured, and it becomes a simple matter to extract the subjective element of distortion.

Is this good enough? If von Neumann was right, the psychological component isn't separate from how we experience the world at every moment. Maybe the "real" reality is waiting out there for someone who can do a better job finding it. Materialists—who prefer to be called physicalists, since their worldview includes energy as well as matter—insist that no psychological component is needed, but the history of quantum physics points the other way. Schrödinger has been dismissed as a mystic, but he knew, based on empirical evidence, that at the basic level a subatomic particle doesn't behave like a tiny planet but like a smear of possibilities. The observer determines which possibility will undergo a change of state, manifesting as an object that can be measured.

So the best answer to the mystery of "Where did time come from?" turns out to be a human answer. We didn't have to be present at the big bang for it to have a psychological component. The only version of the big bang anyone will ever know is the story told by human beings using our mind and brain. The same mechanism is producing reality at this very moment. Therefore, the mystery of time exists before our eyes. Without a human answer, it will remain a riddle forever.

In this chapter we've given you a preview of the benefits of a human universe where time is on your side because you participate in creating it. At the moment, however, physics is still struggling to keep objective time intact, preserving it as the only "real time" that science has to worry about. But what if the only real time is the present moment? That would bring down the wall dividing personal time and objective time. Once that happens, everyday life could be transformed into eternal life, here and now. This startling possibility makes the mystery of time important to everyone. Each of us creates a unique relationship with time, and yet our source is timeless. If we can look past the illusion created by clocks, the race against time comes to an end, and the fear of death is erased once and for all.

WHAT IS THE UNIVERSE MADE OF?

The universe has been putting on a striptease act for a long time. One by one it has shed the veils that cover up the truth about nature. At first the strip was boringly slow. The audience had to wait centuries before the first veil, which was the idea of a solid atom, came off. The atom is an ancient idea, dating back to Democritus and his followers. Those philosophers in ancient Greece couldn't see an atom—neither can we, more than two thousand years later—but they reasoned that if you slice up an object, any object, eventually you'd arrive at a tiny piece that couldn't be cut any smaller. The word *atom* comes from two Greek words that mean "not" and "cut."

The striptease would have gone much faster if someone could have found a way to prove that atoms exist, but they couldn't. Therefore, if you asked what the universe is made of, the answers you got back were all theory and no action. But it was certain that some kind of smallest unit must exist. Peeling away the veils moved incredibly fast starting in the eighteenth century, when experimenters actually began to experiment, and the behavior of chemical reactions gave the first clues that single, whole atoms were reacting with one another. Skip ahead to the twentieth century, when evidence was found for electrons, radiation, the nucleus, subatomic particles, and so on. One by one the building blocks of the atom were discovered. The universe could hide behind modesty no more.

So the audience was shocked when the last veil was dropped and behold, the dancer wasn't there! If you keep slicing a loaf of bread into smaller and smaller units, the atom vanishes into the quantum vacuum. Something turns into nothing, as we've already seen. But there's a subversive side to this striptease. Once the dancer vanishes, we are left with thinking about the universe rather than actually seeing it. Somehow we're back at square one with the ancient Greeks, relying on logic and speculation instead of provable facts.

Right now, outside public view, there is a "battle for the heart and soul of physics" going on, to borrow a phrase from the prominent journal *Nature.* Two highly respected physicists, George Ellis and Joe Silk, wrote an article in 2014 that raises alarms over just this problem of pure thinking replacing data and facts. Can pure thinking be called science, which for five hundred years has pursued the truth through measurements and experiments? Once you get down to nothingness, the zero point of the universe, the possibility of doing experiments comes to an end. How bothered should we be?

Here's an analogy from everyday life. See yourself about to cross the street at a busy city intersection. In front of you is the Walk/Don't Walk sign. Cars are pulling up to the intersection constantly, and some are turning right on red. Your object is to cross the street without getting hit by a car. To make this a real challenge, you must wear blinders, the kind that horses wear when they pull carriages in Central Park, so you can only look straight ahead.

What is your strategy for not getting hit? Your line of vision is very narrow, and all you really have to work with are clues. This is much the same as a physicist trying to look into a black hole or before the big bang or inside the quantum vacuum. For you, the clues turn out to be quite useful. You can use your hearing to listen for cars. You can see when the Walk sign is on. There are other pedestrians on the corner; you can observe them and step off the curb when they do. This gives you a pretty good idea of when it's safe to cross the street. But you don't actually know. The probability is high that you won't get hit by a car, and that's the most you can say.

If you want to see the reality that lies inside a black hole, you can't. You can only figure out the probabilities based on various clues. The same holds true for almost every mystery we are covering in this book. Science has gotten to the point where things are either too small, too big, too far away, or too inaccessible to the most powerful instruments in the world. If you take the tiniest subatomic particle that the largest accelerators, costing billions of dollars, can blast out of the quantum field, the very smallest particles—or whatever they turn out to be—are still 10 million billion times smaller than any accelerator can detect.

Which brings us to a fork in the road. One sign says, *This way to more thinking*, another sign says *Dead end*. Science hates dead ends, so physics keeps diving into deeper and deeper thinking. One camp keeps faith with the time-tested practice of doing experiments and campaigns to build ever bigger particle accelerators—even though, by some calculations, the energy needed for such a gigantic machine would equal all the electricity in all the power grids on Earth. Another camp abandons experiments and opts for pure thinking—the old Greek way—in the hopes that nature will one day offer new evidence that we can't see right now.

Sherlock Holmes and Albert Einstein have one thing in common: they believed in logic. Einstein had total faith in the logic behind relativity. He once said, only half in jest, that if his theory had proved incorrect, "[t]hen I would have felt sorry for the dear Lord." It's strange to think that if you hold a loaf of bread in your hand and ask, "What is this made of?" the ultimate answer is "Nothing, but we have lots of good ideas about that." Such is the present situation when pursuing the mystery of what the universe is made of. There has to be a better way.

GRASPING THE MYSTERY

When a problem arises where the evidence is kept out of sight, it's known in science as a black box problem. For example, imagine that new cars roll off the assembly line with their hoods sealed shut. No one can see the car's engine—it's in a black box—but you can still tell a lot about how the car runs. One by one, facts can be amassed. When the car suddenly stops, for example, you will eventually discover that it needs gas. Because the dashboard lights up, you deduce that the engine involves electricity in some way.

Black boxes are fun and frustrating at the same time, and scientists tend to love them. But until you can open the hood, you will never know how a car engine actually works. It's very disturbing, then, to realize that the universe itself is the ultimate black box. If a physicist sets out to understand what the universe is made of, everything seems to be on the table. The laws of nature are well understood, as are the properties of matter and energy. The standard model of quantum field theory can account for every fundamental force except gravity. Even though gravity is a stubborn holdout, tiny increments of progress continue to be made (at the moment the two leading rivals are known as loop quantum gravity and superstring quantum gravity, both highly esoteric), and everyone keeps murmuring that slow and steady wins the race.

Unless it's all reached a dead end. The infant universe was cooked up where no one can go, or even name the raw materials that were used. As Ruth Kastner, an accomplished philosopher of science, has commented, the material universe is like the Cheshire Cat in *Alice's Adventures in Wonderland*. Its body has faded away, leaving only a faint grin hanging in the air. Physics studies the grin in an attempt to describe the cat. Is this a futile enterprise? The Cheshire Cat metaphor originated with the work of far-seeing physicist John Archibald Wheeler to describe the collapse of matter into a black hole. Einstein had a witty way of putting it: "Before my theory, people thought that if you removed all the matter from the universe, you would be left with empty space. My theory says that if you remove all the matter, space disappears, too!" When you consider that a black hole devours literally the entire structure of physical reality, it's easy to look upon even a huge cluster of whirling galaxies as nothing more than the cat's grin.

Physics wants to find a single explanation of reality. But there's no getting past the fork in the road. One way leads to a universe where matter is substantial, reliable, and well understood. Quantum physics more or less abolished this as a viable route to reality, even though large numbers of working scientists still choose this path. They have their reasons, which we'll examine. The other way leads to a total rethinking of the universe, based on the fact that material existence is an illusion. The dilemma is like Robert Frost's famous poem that begins, "Two roads diverged in a yellow wood, / And sorry I could not travel both . . ."

Most of the unsettled arguments in quantum theory turn on which road you decide to take. Pure thinking or new data? As in Frost's poem, the most frustrating aspect is that you'll never know what happens on the road not taken.

PRYING OPEN THE BLACK BOX

Cosmologists accept that the visible universe constitutes only a fraction of the matter and energy unleashed by the big bang. The vast bulk of creation almost instantly disappeared, but this didn't remove dark matter and energy from the equation. For example, empty space isn't empty but contains huge amounts of untapped energy at the quantum level. The exact amount of energy has been calculated, but it turns out, on the evidence of how fast the universe is expanding, the numbers are way, way off. As subatomic particles "foam" up from the vacuum, the forces involved require enormous amounts of energy. The density of energy in a cubic centimeter of empty space is expressed as a number called the cosmological constant.

Unfortunately, this number turns out to be off by 120 orders of magnitude (10 followed by 120 zeros). Empty space is far emptier than quantum theory would have it. Somehow, it is supposed, all the forces that should be roiling inside the vacuum state cancel each other out. More than one physicist has called this perfect cancellation "magical." At best, what's happening is due to dark energy and its effects on the galaxies, but dark energy is high on the list of things that so far at least can't be experimented on.

If it turns out that the hidden side of creation is actually in control of the expanding universe, we confront possibilities that defy the accepted view of the laws of nature (the standard model). In a nutshell, when solid, reliable matter vanished, so did the concept of "matter." This will turn out to be tremendously important if all the things we take for granted about physical objects—the heaviness of rock, the sweetness of sugar, the brightness of a diamond—are created in the human mind. This would imply that the whole universe is created in the human mind—but we're not there yet.

To give an idea of the gap, no one really knows why the physical universe exists in the first place. During the big bang, energy was wildly active, producing a "shakeup" of space-time. The calculations of physics can't tell us why such violent agitation didn't doom matter to be torn apart. If primordial matter was shaken as much as the equations say it was, either the infant cosmos would have collapsed in on itself by the tremendous force of condensed gravity (as in a black hole) or the surviving universe would have been pure energy. Yet it's obvious that matter did come into existence; therefore, the equations must be tinkered with until they fit how things are. This tinkering can look a lot like fudging the numbers.

Reality is obviously more than physical, and trying to squeeze quantum "stuff" into a physical box isn't what reality is telling us to do. Yet a belief in physicality remains part of most scientists' DNA. They point to the success of the standard model and promise that all the remaining gaps will soon get filled in. "We're almost there" fuels optimism. Nonphysical explanations for the universe would go back to the starting point, based on accepting that "matter" is a worn-out concept. Given a choice between "We're almost there" and "We haven't even started," most scientists pick the former without question.

WHAT WE SEE

Before radically challenging the physicalist position, credit must be given to the knowledge it accumulated. It's an impressive achievement, all of it based on the maxim "Seeing is believing." There's certainly a lot to see. Within 14 billion light-years or so (the actual universe may be much larger), there are probably 80 billion galaxies, which astronomers classify as large and small, spiral, elliptical, or irregular in shape, "normal" (showing no major activity in their centers), or "active" (exploding with vast amounts of energy and matter coming out of their centers).

Within a typical galaxy like our own Milky Way, a large spiral type, there are as many as 200 to 400 billion stars. Most of these are called red dwarves; are small, faint, and red in color; and last for tens of billions of years. The stars we see in the night sky are much brighter, and they have whitish or bluish colors. These bright stars can be seen from much farther away, but what we see doesn't reflect their true distribution. A high percentage of stars other than red dwarves are like our own sun, and many are now shown to be surrounded by planets. As we saw, if a percentage of these planets harbor the right conditions for life, then the camp that believes in randomness has an advantage over the anthropic camp, with its belief that life on Earth is special.*

In total, the universe contains as many stars as 1 followed by 23 zeros, or 100 sextillion. A staggering number, but by no means the most staggering. A vast amount of luminous matter, in the form of stars, lights up the galaxies. Even though there are more stars than grains of sand on Earth, they only make up 10 percent of the total mass in the observable universe. Calculating the total number of protons and electrons making up regular atomic matter, one comes up with 1 followed by 80 zeros, or 100 thousand trillion sextillion sextillion sextillion atoms! This is equivalent to 25 million sextillion Earths.

Here the visible trail dwindles away, because all this luminous matter accounts for approximately 4 percent of the "stuff" in the universe. Most of it, 96 percent or so, is "dark" and therefore unseen and unknown. But at least we have a credible inventory of the cosmos, as produced by NASA's Wilkinson Microwave Probe (WMAP): 4.6 percent regular matter, 24 percent dark matter, and 71.4 percent dark energy. Most of the universe is at the very least quite exotic. Quite a black box indeed.

As things stand, dark matter and energy are surmises formulated by painstaking, elaborate lines of reasoning—their actual existence is several steps removed from "seeing is believing." Some skeptics warn that physics is flirting with fantasy. Imagine

* To date, NASA's planet-hunting Kepler spacecraft has spotted 1,000 potential Earths in deep space. As we were writing this book, a new candidate, Kepler 452b, was added to the list. Situated 1,400 light-years away, making it one of the closest possibilities, the size of Kepler 452b and its distance from the star it orbits fall into the "Goldilocks" zone of perhaps being not too hot or too cold to sustain oceans and be suitable for life.

that you are looking around the animal kingdom and see horses galloping through the open plains. Turning your gaze, you also see a one-horned sea mammal called the narwhal. Do these visible facts allow you to reason that unicorns are real, with the body of a horse and the horn of the narwhal? Our modern answer is no, but in the Middle Ages there wasn't such a strict divide between the real and the mythical. Cosmology is currently saddled with a full menagerie of mythical creatures, from quarks and superstrings to the multiverse, created through mathematical inference alone.

Dark matter is a prime example of real-by-inference. First, dark matter is inferred from the speeded-up rotation of the stars in a typical galaxy. The stars are being pulled around by the gravitational force of some outside mass faster than physics can account for. (NASA uses gravity in the same way when it steers a space probe close to a huge planet like Jupiter or Saturn so that the planet's gravity can serve as a slingshot, speeding up the probe as it whips past.) As normally measured, the typical galaxy doesn't contain enough mass to explain the observed rotation, nor does the known universe.

Second, most galaxies are found in clusters of various sizes. Some are small, containing only a few galaxies, while others are massive, containing tens of thousands of galaxies and emitting vast amounts of X-rays. These giant clusters also seem to contain more mass than is counted up, either in stars or the gaseous material inside the cluster, which is only observed through X-rays. By inference, more matter must be contained somewhere inside the cluster. Finally, when distant background galaxies are observed as their light passes through a nearer cluster of galaxies (such as the Bullet Cluster), the bending of its light due to the gravitational field inside the near cluster—acting as a gravitational lens—indicates that a lot more dark matter resides inside the cluster. These three pieces of evidence are in agreement, based on the same variable, gravity. They result in precise numerical predictions that have been confirmed. The inferences being drawn aren't weak, but they aren't sufficient either.

To illustrate, imagine that you are in a windowless room that's rotating like a star. You can feel the centrifugal force as you are bounced against the walls, and you infer that something is pulling on the room from the outside. That's a strong inference, but you can see its limitations—to describe where the external force is coming from (a tornado, an angry elephant, a giant playing with his toys?), nothing can be said realistically with only an inference to go on, despite the finest calculations from inside the room telling you how strong the force is.

WHEN DARKNESS RULES

Since darkness appears to be the rule in creation, solving the mystery of what the universe is made of must begin there—and almost immediately gets stymied. Most cosmologists currently believe that dark matter is "cold," meaning that within a year after the big bang, its particles were moving slowly in relation to the speed of light. (As you've come to expect, these particles are just a matter of conjecture at this point.) It's also been proposed that dark matter may come in three types: hot, warm, and cold. For example, subatomic particles known as neutrinos have been nominated to form hot dark matter, bringing it closest to the realm of ordinary matter. Warm dark matter is thought to exist as "brown dwarfs," objects that are too small to light up via thermonuclear reactions the way ordinary stars do.

On a more solid footing, the consensus today holds that cold dark matter is composed of weakly interacting massive particles (WIMPS), which are heavy and slow-moving. The aptly named WIMPS interact via gravitation and the weak force alone. They would be completely hidden were it not for their distribution over the whole universe and the large proportion of total matter they constitute, exerting a powerful gravitational force.

Dark energy is considerably more exotic and seems to be vastly more present. Whereas dark matter, although unseen, still influences the visible universe via its gravitational pull, dark energy acts as antigravity, pulling the universe apart at large scales (e.g., beyond the scale of galaxies and clusters of galaxies). How this actually takes place, and providing a theoretical explanation, is no small mystery. Even for it to exist requires precise measurements of how fast the galaxies are accelerating away from one another. According to how many stars you take into account—the key ones are very distant supernovas—the value for dark energy shifts considerably. Some skeptics challenge whether the galaxies are accelerating at all, which would undercut dark energy completely. But cold dark matter with dark energy is currently taken to be the standard model of cosmology. We supposedly inhabit a flat universe, dominated by dark energy, with smaller amounts of dark mass and even smaller amounts of luminous, or ordinary, matter.

From another viewpoint entirely, darkness could be a case of how we observe the universe rather than what it really is. The giant particle accelerators that blast subatomic particles into view operate on the tiniest scale, mere billionths of a meter and billionths of a second. Is that kind of observation compatible with the effect of dark matter, which operates on the largest scale, billions of light-years in size? Before anyone can answer yes or no, one has to challenge if what we see today is the same as what existed long ago. Almost certainly it isn't. The acceleration that is making the galaxies fly apart faster and faster began very late in the game, roughly 6 billion years ago. Before then, cosmologists believe that the expansion was actually decelerating. That's because dark matter and dark energy evolve differently in an expanding universe. When the early universe doubled in volume, dark matter density was halved, but dark energy density remained (and remains) constant. When the balance tipped in favor of dark energy, deceleration turned to acceleration.

The "We haven't even started yet" camp is bolstered by gaps in the standard model. What would it take for totally new thinking to take hold? The journey begins with the psychological aspect of reality, which von Neumann called essential. Seconding him are an array of eminent physicists from the beginning discoveries the quantum era. Max Planck was adamant that reality at bottom involves consciousness. As he put it, "All matter originates and exists only by virtue of a force. We must assume behind this force the existence of a conscious and intelligent mind. This mind is the matrix of all matter."

This means that lumps of matter are no longer floating "out there" like snowflakes that fall from the sky and collect on your coat collar, but rather, matter is embraced in the same matrix that holds thoughts and dreams. Planck's belief that mind is even more basic than matter is expressed with total clarity here: "I regard consciousness as fundamental. I regard matter as derivative from consciousness. . . . Everything that we talk about, everything that we regard as existing, postulates consciousness."

If you're looking for totally new thinking, it's been around awhile. What was lacking was acceptance, so let's build some.

REALITY IS A MIND GAME

Pioneers are bold almost by definition. But what made Planck join Schrödinger in his staunch belief that the universe is mindlike? It goes back to a fact almost too basic to need stating, namely, that everything we experience is an experience. Does this actually tell us anything? Burning your tongue on hot coffee is obviously an experience, and so is building the New Horizons space probe, launching it with a huge missile so that it travels at 36,000 miles per hour through space (boosted to 47,000 mph as it receives a boost swinging around Jupiter), waiting nine years for it to journey almost 6 billion miles to Pluto, and then sending up a cheer, as astronomers did on July 14, 2015, when New Horizons sent back the first close-up photos of the last major body in the solar system.

Burning your tongue and photographing Pluto stand on equal footing as experiences; and doing any kind of science is also an experience. So Planck was asserting that this fact counts—all the time and very deeply. If you can level things as different as the smell of a rose, the blast of a volcanic explosion, a Shakespeare sonnet, and a space probe, then the "matrix" of reality is no longer physical. This offers a tremendous advantage when you reach the dead end that physical "stuff" has reached. The simplicity of turning to a totally new paradigm is that darkness no longer has to be considered alien. The matrix has no trouble including it, because all the stuff in the universe has become mind-stuff.

Here the physicalists shove their oar in. Making solid objects vanish is child's play compared with bringing them back again. How does mind-stuff, which has no mass or energy, manage to create mass and energy? The matrix that Planck calls consciousness is nothing more, the physicalists might claim, than the universe with all its mysteries unsolved. Sticking on the tag "consciousness" doesn't really produce any answers. (This skeptical attitude has been paraphrased as "What is matter? Never mind. What is mind? Doesn't matter.") In the spirit of fairness, the two camps face equal but opposite difficulties. One must show how the material universe developed the phenomenon of mind, while the other must show how the cosmic mind manufactured matter. At first blush, we're back in the big muddy of theology, which failed to answer how God did either one.

THE OBSERVER PROBLEM POKES ITS HEAD UP

John von Neumann, having included a psychological component in his version of quantum mechanics, seems to have a foot in both camps. But it's a shaky place to stand. Let's say that he was right and reality cannot be separated from personal experience. This doesn't explain how an experience dips into the quantum level. There's no doubt that subjectivity is a powerful force for altering reality. As humorist Garrison Keillor says on his popular radio show, *Prairie Home Companion*, "Well, that's the news from Lake Wobegon, where all the women are strong, all the men are good-looking, and all the children are above average." That's subjectivity overriding reality. But it's another thing to hold that subjectivity creates reality.

The problem becomes easier if we stop seeing subjectivity as the opposite of objectivity. They are actually merged into each other. The reason we know this is because the subjective side of experience can't be isolated or subtracted. In other words, when everything is an experience—and everything is—subjectivity must always be present.

Naturally, the physicalist camp resists this claim quite strongly. For a century this great bone of contention has been known as the observer problem. Before it can measure something, science must first observe it. In the classical world, there was no problem observing whatever lay before us: tadpoles, the rings of Saturn, or light being refracted through a prism. One experimenter could leave the room, and no matter who took his place, the observation would remain the same.

The observer is only a problem if the very act of looking creates a change in the thing you are looking at. In the human world we encounter this all the time. If someone gazes at you with love in their eyes, you are very likely to change, and you will change again if the look becomes indifferent or hostile instead. This change can extend very deep, to physical reactions in your body. If you blush or your heart beats faster, the chemistry of your physiology is reacting to a mere look. What makes the observer problem unique in quantum physics is that the act of observation can be enough to bring particles into existence in time and space. This is technically known as the collapse of the wave function, meaning that a probability wave, which is invisible and extends infinitely in all directions, changes state, and suddenly a particle is visible.

One of the basics in quantum mechanics is that a quantum (for instance, a photon or electron) can behave either like a wave or a particle—no one disputes this. What is disputed is whether the simple act of observation causes the wave function to collapse. On the physicalist side, things are things, period, and claiming that an observer causes a particle to emerge from the quantum field is mysticism, not physics. But the most widely accepted version of quantum mechanics, the Copenhagen interpretation (so named after the work done at the Copenhagen Institute by Danish physicist Niels Bohr), places the observer at the crossroads between wave and particle.

This still leaves open the mechanism that allows the act of looking to affect physical matter. Something must be going on under the table, as it were. Observer A looks at Object B with the intention of measuring something about it, such as its mass, position, momentum, and so on. The instant this intention is specified, the object complies—that's the under-the-table part. No one has an accepted explanation for it. Heisenberg described this in the most definite terms: "What we observe is not nature itself, but nature exposed to our method of questioning." The observer cannot be separated from the observed, because nature gives us what we want to look for. The whole universe, it seems, is like Lake Wobegon.

Now let's extend the observer problem, which in the Copenhagen interpretation becomes the observer effect, to the mystery of what the universe is made of. If, as Heisenberg said, "atoms or elementary particles are themselves not real," then asking what the universe is made of turns out to be the wrong question. We are trying to squeeze juice out of an illusion, and it won't work. The universe is made of *what we want it to show us*. Physicalists roll their eyes when they hear such an idea, but certain facts are undeniable. No one has ever seen the wave function collapse—it's not an observable event—whereas calculating the behavior of matter in terms of uncertainty and probabilities has proved spectacularly successful. Quantum objects defy commonsense rules of cause and effect.

Put these facts together, and the picture that emerges isn't a cosmos full of "stuff" but a cosmos full of possibilities mysteriously turning into "stuff"—the transformation is more real than the physical appearance we take for granted. To date there is no better answer to "What is the universe made of?" Even a grumbling physicalist has to concede that the collapse of the wave function is a transformation. Pulling a rabbit out of a hat is an illusion; pulling a photon out of the field is real.

Unfortunately for the Copenhagen interpretation (and all of modern physics, no matter which interpretation one favors), the road stops here. An observer in the laboratory may affect the behavior of a photon, but this is miles away from everyday life. Can looking at the whole universe, its stars and galaxies, or looking at trees, clouds, and mountains actually transform them? The notion sounds preposterous at this point, but in fact, this is the basic claim of the human universe. We aren't there yet. To get around the roadblock, we'll have to prove that mind isn't just one factor in the cosmos but the factor that underlies how everything in creation behaves. That challenge is looming ever closer, one mystery at a time.

IS THERE DESIGN IN THE UNIVERSE?

Are we living inside a universe with a grand design? This was a hot-button question long before "intelligent design" sent off alarm bells among the scientific community. Intelligent design is based on trusting the book of Genesis, but if you loosen your criteria and ask, "Is God playing any part in creation?" the same firestorm erupts. Science is anti-design because of its stance on religion (keep it out of the lab), politics (don't let churches interfere with government funding), and rationality (there are no data suggesting a grand design driven by God or the gods).

A random universe excludes the notion of design. If every event happens by chance, from the emergence of a subatomic particle to the big bang, there's no need for a designer to oversee how the cosmos turned out. So why is there any mystery to solve? Because our minds are caught between two worldviews—it's like being trapped in an elevator stuck between two floors. In Rudyard Kipling's children's story "How the Leopard Got Its Spots," the spots were painted on by an Ethiopian hunter so that leopards would blend into the "speckly, patchy-blotchy shadows." Modern science agrees: it turns out that cats who hunt in the dark or in dappled forest light are much more likely to have spots or stripes, because these evolved to help the animals hide and hunt for food. Cats that hunt out in the open are more likely to have plain, undecorated coats. (There's always an exception to the rule, so we get the cheetah, which runs after its prey out in the open but is also spotted.)

It would seem that Kipling and an evolutionary biologist might have come up with the same answer—only they didn't. In place of "Ethiopian hunter," put the words *God* or *Mother Nature* or whatever "designer" you like; in the form of a whimsical tale for children, Kipling adheres to the worldview that gives the leopard its spots *for a reason*, and this reason is known in advance—camouflage. This worldview doesn't specifically need God, just a creative reason for leopards to have spots. The Ethiopian hunter didn't paint the leopard bright orange, because that would have defeated the whole point.

Science puts the reason afterward, as the effect, not the cause. Leopards got their spots at random, due to the interaction of two specific chemicals known as morphogens. These chemicals create all patterns, including the ridges your tongue can feel on the roof of your mouth. Through a random mutation involving morphogens and how they interact, spots emerged on a cat long, long ago, and it turned out to work as camouflage. The animal doesn't know that it's camouflaged; it doesn't know anything about how it looks. The only thing that matters in Darwinism is survival, and a cat with spots survives better by being a better hunter in dappled light. (The patterns of spots and stripes on cats in the wild are random also, and their arrangement was predicted using a computer model by World War II British codebreaker Alan Turing.)

Why, then, are we stuck between worldviews like an elevator stuck between floors? Because in our minds, there's a reason for leopards to have spots, just as Kipling said, but at the same time we accept the mechanism behind the spots, just as science says. It is very hard to get the human mind to accept that absolutely everything in nature is meaningless, but that's what Darwinism, the big bang, cosmic inflation, and the formation of the solar system are all about—stripping creation of human notions like purpose and meaning.

Scientists hate the word *design* because it feels like a sneak attack from a worldview they thought was extinct. But if you forget the current hot-button climate of opinion, the words *design*, *pattern*, *structure*, and *form* are synonyms. There is no rational reason why "design" should be considered especially radioactive.

But we have to be realistic. Words have histories, and the history of "design" is repugnant to many scientists because of its association with creationism. The creationist campaign updates the book of Genesis by claiming that science supports the notion of intelligent design. Alarmists on the opposite side view this as a threat to the integrity of science. In reality, intelligent design has appealed mostly to the faithful and to the mass media, who knew an entertaining story when they saw one.

The courts have thrown out any attempt to give creationism equal time in school curriculums with science (although a few exceptions linger, unfortunately). It would seem foolhardy to plow this field again. But that stuck elevator won't budge. Looking around at nature, we see design everywhere. Is this just a trick of the mind? No one catches bears and frogs staring with wonder at a rainbow. For them, there is no beautiful iridescent arc, in fact no pattern at all. Accounting for the beauty of a rainbow may be a red herring. Perhaps we should be asking a totally cold-blooded question: Is *anything* in the universe there by design?

GRASPING THE MYSTERY

Despite their belief in randomness, scientists regularly refer to the structure of the atom. Spiral nebulae form a recognizable pattern that one can harmlessly call a design, and with this in mind, the whole messy issue of design-pattern-form-structure can be clarified as follows: The universe owes its existence to the emergence of order from chaos. The wrestling match between form and the formless is still with us throughout the universe. Modern physics is based on random processes devoid of purpose and meaning. (We don't ask a question like "What does gravity *mean* on Jupiter?") And yet human life, including the pursuit of science, has purpose and meaning. Where did those come from?

Without a doubt, the language of mathematics exhibits every quality of design: balance, harmony, symmetry, and some would say beauty. In Chinese calligraphy, the ability to draw a perfect circle with one stroke of the brush is the mark of a master, and art connoisseurs see beauty in the achievement. Electrons, at least for the lowest orbits, travel in a perfect circle around the nucleus of an atom. Isn't that a beautiful design, too? The following are all examples of helixes, or spirals, in nature: the shell of a chambered nautilus, the pattern of seeds in a sunflower, and the structure of DNA. Which ones qualify as a design—some, all, or none?

A science that depends totally on randomness to explain the universe falls far short. Inside the rational activity of science there is still much to argue over, because intelligence and design are tangled in the same ball of yarn that makes the universe so mysterious. We'll try to unravel the snarl without any agenda, but this will take the exposure of some hidden agendas along the way.

We accept Bohr and Heisenberg's insight, which was quite brilliant, that nature displays the properties an observer happens to be looking for. This notion certainly pertains to design. Nothing about a rose—the rich crimson color, the velvety texture or sharp thorn, the sumptuous fragrance—exists without an observer. Yet your mind can hold a gorgeous red rose in full bloom because the human brain transforms, or translates, raw data into sight, sound, touch, taste, and smell. There isn't even any light in the world without someone to see it, because photons have no brightness on their own. Deep within the pitch-black recesses of the visual cortex, purely chemical and electrical impulses traveling along the optic nerve are transformed into light.

The fact that the brain is totally dark while the world is full of light could be called the mystery of mysteries, and we're not quite ready to tackle that one yet. For the moment, we'll stay with the tie that binds observer and observed. If it takes the brain to process the raw material of nature and turn it into a beautiful red rose, is the same processing also creating design? Clearly the answer is yes. When a caterpillar is chewing on a rose, it can destroy its beauty in an hour, but the beauty of the rose that the caterpillar takes away was put there by human beings. To an insect that preys on roses, a flower is just food.

It's not really the brain that creates beauty but the mind. Someone who is violently allergic to roses might consider them too great a nuisance to be beautiful. Such a person presumably has the same brain mechanisms as Pierre-Joseph Redouté, a famous painter of roses during the time of Napoleon, but their mind-set isn't the same. And if roses are beautiful only because the human mind finds beauty in them, does the same hold true for the entire cosmos? Putting the question this way seems innocent enough, but it has explosive implications.

One camp that gets particularly agitated is known as naïve realism. In scientific arguments, the naïve realists are the great defenders of common sense, using reality as it is given to buttress their position (the word *naïve* isn't meant to be pejorative; it's just the opposite of overthinking).

Here are two givens that apply to the human brain, for example:

Every thought is accompanied by the firing of neurons.

Many thoughts contain information, such as 1 + 1 = 2.

No one would dispute these facts and, according to naïve realists, observing neural activity on a brain scan is enough to tell you that the brain creates the mind, that the brain is basically "a computer made of meat," to use an unlovely description that is popular in the field of artificial intelligence, and that every enigma posed by the brain can be solved be examining its physical structure and operation.

At a guess, 90 percent of neuroscientists and an even higher percentage of researchers in artificial intelligence (AI) believe in these ideas; therefore you can see the power of naïve realism. Yet from another angle, AI is making an obvious mistake. When you ask your computer to translate a page of German into English, a translator program can do it almost instantly. Does this mean that your computer knows German? Of course not. The artificial imitation of thinking isn't the real thing. The translator program does its job by matching words and phrases to a dictionary. Someone who knows German doesn't do this at all. Thinking requires a mind, period. Even though two facts about the brain are true, to say that the brain creates the mind and that computers and brains are the same isn't automatically true. These are simply assumptions, and naïve realism is full of other assumptions that get accepted without being examined. Unexamined assumptions make it harder to untangle the thorny mystery of design. But the assumptions are still there, even if swept under the carpet. Because naïve realism only looks at reality-as-given, it discounts the role that mind plays. Many AI experts believe that a translator program turning *guten Morgen* into *good morning* is the equivalent of performing a mental action and therefore, the resemblance to a human mind is proved. But if mind is actually the leading player in the universe, naïve realism is totally off-base, no matter how many scientists believe in it.

The mind-like behavior of the cosmos has been cropping up frequently in our discussion. Now we're ready to confront its greatest challenger, which is randomness. Randomness implies "having no purpose." Yet the two are not the same, as we will show in relation to quantum activity. If the universe is totally random and with no purpose, all possibility of finding design will fail. On the other hand, if there is some way to make peace with randomness, as quantum theory attempts to do, the cosmos inches closer to acting like a mind—not just that, but a human mind. As you sit in a chair dangling your feet, they move more or less randomly. When you get up to go to the refrigerator for a snack, your feet move purposefully. This gives us the simplest and yet most profound clue. Randomness and design cooperate with each other, in nature, in our bodies, in our thoughts. Let's see if this insight is enough to pry open the tight grip that pure chance holds on the practice of science.

TAKING A CHANCE ON CHANCE

The great god of randomness had a modest beginning when physicists wanted to explain basic phenomena, such as how gas molecules behave. If you watch motes of dust dancing in a sunbeam, their motion is random, which poses a scientific problem. How can you predict where any one dust mote will be located in the future? Is it impossible or merely very, very difficult? With regard to gases, it was assumed that the overall behavior of gas molecules, which teem in far greater numbers than dust, can be understood if the individual motion of each particle is taken to be random, which makes their specific locations in space indeterminate. (This is a good assumption for any large collection of particles.)

Even though the microscopic properties of individual molecules are unknown, the average macroscopic properties of the whole collection of molecules can be easily defined. You simply sum up the average motion of each molecule. The properties of dancing gas molecules are covered in a branch of physics called thermodynamics, because the heat, or thermal state, of a gas causes it to move faster as the temperature rises (which is why boiling water bubbles with rapid motion—heat causes the water molecules to turn to steam, which is a much more agitated state). The average motion can be accurately used even though a particular molecule's motion is unknown. So, by knowing only one parameter, namely temperature, randomness can be handled as a practical matter.

How far can one legitimately take this kind of averaging? That's a question that doesn't get asked enough. Averaging can lose as much knowledge as it gains. If you are in a helicopter flying over a busy freeway, you can't predict which exit a specific car will take. Using a statistical average, a reliable number emerges as it applies to all the traffic on the road, but you've totally overlooked the most important thing: randomness in this case is a total illusion. Each driver knows where he's going and takes the exit he needs. Drivers aren't making random decisions, even though from the outside their activity may look random. This distinction leads in various directions. You can't predict the next thought that will enter your head, but calling thoughts totally random is wildly off the mark.

When you're thinking about what to have for dinner tonight, you aren't engaging in random musings—there's a purpose to your thinking. Yet we all daydream, and incidental thoughts do float like mental lint through the mind. This tells us that making peace with random chance isn't just an arcane issue or some kind of intellectual game. There are many ways that randomness can fool us. A lot depends on who is observing what. Imagine an ant crawling across a painter's palette while the artist is at work. The ant scurries this way and that as the tip of a paintbrush jabs randomly at red, blue, green—it has no idea which color will be jabbed at next—while, from the artist's perspective, the randomness is the illusion, as each tiny brushstroke serves a purpose in the artistic creation.

Pure randomness, if you aren't totally wedded to it, never tells the whole story. Naïve realists, seeing dust motes dancing in a sunbeam and gas molecules bouncing off one another, have overtaxed the usefulness of the observation and willfully ignore the possibility, so brilliantly intuited by Heisenberg, that nature gives every observer what he's looking for.

Separating order from chaos was relatively simple in classical physics, but it became much murkier in the quantum era, when it was proposed that particles behave randomly *in principle.* It's not practical to determine the position of every air molecule in a room, but in classical physics, using a mythical supercomputer with unlimited speed and memory, one could calculate where each one is located and where it will be an hour from now.

The same isn't true of subatomic particles in the quantum universe. The uncertainty principle assures us that particles have no well-determined position and motion, only probable ones. What is the probability that all the oxygen atoms in a room will congregate in one corner? Practically speaking, the odds are zero. But a beautiful calculation known as Schrödinger's equation can deliver the exact probability, to many decimal places, of such an event, however minuscule. We no longer have to use averaging. Randomness has found a much more precise and elegant way to be calculated.

But this success doesn't mean that the same headway has been made in balancing order and chaos. How one translates into the other is often inexplicable. Even the most precise predictions have their flaws. Imagine a mechanic's garage where they can measure the tread on your tires and predict within half a mile when a tire will blow out. That would be impressive, but the prediction has nothing to say about what road you will be on when the tire blows out, why you chose the road, and what your destination is. If the mechanic shrugs his shoulders and says, "Those things don't matter to me. They're out of my hands," you'd agree. But the road that molecules, atoms, and subatomic particles take, and the destination they are headed for, can't be dismissed. In your bloodstream, whether a molecule of cholesterol winds up clogging a coronary artery or passing harmlessly out of the body can be a matter of life and death.

Because of their physicalist beliefs, many scientists keep averaging out difficult problems as if that's the best (or only) way to handle randomness. A striking example is evolution. Looking at an elephant, you can see that its snake-like trunk and sail-like ears are unique. The elephant evolved to have them, and according to Darwinian theory, having just this kind of trunk and that kind of ears enabled the first elephants to survive better. New adaptations begin at the genetic level with a mutation that hasn't been seen before. Mutations occur at random, so standard evolution says, and they must be passed on to later generations in order to become permanent. If a single pink elephant appeared millions of years ago, we'll never know, because that genetic mutation failed to be passed on.

How did the first elephant with a long trunk gain its survival advantage? Impossible to say. It's not even clear that one elephant did gain an advantage—but the whole species eventually did. Without knowing anything that happened to the individual elephant, a kind of averaging out is done by looking at all elephants. In other words, evolutionary thinkers are treating creatures with very complex lives as if they were a collection of gas molecules. On the face of it, this seems like fudging. Animal existence is filled with sudden necessities (like drought or a disease epidemic), unique events, unknown challenges, and so on. Every step of the way, each lion, chimpanzee, or otter makes choices.

Erasing these complexities from the equation in order to get a good group approximation can't be telling us the whole story—perhaps not even the right story. For example, survival of the fittest (a phrase Darwin never used, by the way) can supposedly be reduced to two components: success at getting enough food to eat and the ability to beat out rivals for mating rights. On that basis, gene mutations get passed on. But this picture of constant competition overlooks the fact that in nature cooperation is just as common as competition. Birds flock together, fish swim in schools, and there are myriad other observable populations that live together for safety and shared resources—sometimes seeming to act almost as a single organism. In many marine species, all the males and females gather in one place to scatter clouds of eggs and sperm into the water, acting as one giant mating party with no one excluded. Darwinian evolution has been modified by some theorists to include cooperation, but finding the balance between competitive and cooperative behavior has proved very tricky and controversial.

WHEN CHANCE IS DETHRONED

Let's say that the worship of randomness has seriously faltered, and an old god is ready to topple. How do you balance order and chaos then? If nature is secretly an artist making creative decisions, random events are like the jabbing paintbrush from the ant's perspective, and there are enticing clues indicating that this isn't just a fanciful metaphor. Over and over we've reinforced the message that physicists trust mathematics. The fine-tuning problem opened a rift in the universe being one vast playground of coincidences. In the same vein, some numbers keep reappearing in nature at very small and very large scales.

One kind of design that remains untainted is mathematical. We already discussed, in relation to fine-tuning, how the constants suspiciously match up with one another. You'll recall that Paul Dirac was convinced so many matchups couldn't be merely a long stream of coincidences—he searched for an equation that would defy randomness by finding hidden design.

Mathematical design is one reason some physicists accept that the cosmos has structure and form. One of history's lost lives is that of Euclid, the father of geometry, who made the greatest contribution to mathematics in the ancient world. A Greek who lived in Alexandria in the fourth century BCE under the pharaoh Ptolemy I, Euclid left behind no biography. Anecdotes exist of him drawing lines in the sand as he worked out the rules governing circles, squares, and the other geometric figures that we understand thanks to him. Though the stories are fictions, the most astonishing thing about Euclid, and about the mind of Greek mathematicians generally, is the impulse to reduce nature to neat geometrical patterns.

For centuries scientists continued to look for straight lines, circles, and regular curves, driven by a belief that nature embodied perfection when in fact the patterns in nature are often rough and approximate. The roundest tree trunk that looks like a Greek column from a distance has irregularities in its bark; a ball thrown as straight as possible will have its trajectory bent by wind, air resistance, and gravity. Even a bullet fired as straight as possible is actually describing a complex curve once you take a wider perspective that includes Earth's wobbly rotation on its axis and its lopsided orbit around the sun. After relativity, geometry entered four dimensions, which swept Euclid's neatly geometric two-dimensional shapes off the table, and then the quantum revolution offered an entirely new, exotic mathematics not yet unified with General Relativity.

None of these drastic changes, however, negates the notion of cosmic design. What they eliminate is simple geometric design, the perfect circles, squares, and triangles that were supposed to lie at the heart of nature. Even so, DNA is still a beautiful double spiral; rainbows describe a perfect arc (and from the viewpoint of airplane pilots they are complete circles); baseball pitchers can (and must) calculate what kind of curve—or no curve—the ball will follow on its way to the plate. If nature exhibits these designs in the everyday world but is built from totally random events in the quantum world, a huge disparity has arisen and needs to be resolved.

One possibility, offered by Roger Penrose, is that design exists in a region beyond both worlds, where there is only pure mathematics. Here, Penrose proposes, we encounter immortal qualities that resemble the pure "forms" of Plato. Plato saw these forms as the origin of qualities like beauty, truth, and love. The notion that pure, divine love is the source of all love was very appealing. Linking the divine and the human came naturally to all traditional cultures. Penrose wasn't looking for a divine source for the cosmos, but he sees a purity in mathematics (most mathematicians would agree). More important, if math exists beyond all created things, it stabilizes the constants and anchors reality in a place untouched by nature's chaos, roughness, and irregularity.

Penrose's notion of Platonic forms in the domain of mathematics hasn't been widely accepted. He described these forms in objective terms, far from the subjectivity of love, truth, and beauty. "Platonic existence, as I see it, refers to the existence of an objective external standard that is not dependent upon our individual opinions nor upon our particular culture." Penrose wants to base reality on a kind of perfection that is beyond all change. Even though his life's work was based on mathematics, he recognizes that there is a deeper kinship with Plato, who thought that everything in daily life—oak trees, calico cats, water—had a perfect Form (usually capitalized when referring to specific entities).

Penrose has no objection to extending his theory beyond mathematics. "Such 'existence' could also refer to things other than mathematics, such as to morality or aesthetics. . . . Plato himself would have insisted that there are two other fundamental absolute ideals, namely that of the Beautiful and that of the Good. I am not at all averse to admitting the existence of such ideals." This candid admission works against him with scientists who would apply eternal existence only to numbers, but if you stand back, calling mathematics orderly and balanced isn't starkly different from calling it beautiful and harmonious.

BEAUTY TRANSCENDS A ROUGH-AND-TUMBLE WORLD

Nobel laureate Frank Wilczek has taken the next step and offered a physicist's defense of beauty as a human ideal rooted in reality "out there." His brilliant 2015 book, *A Beautiful Question*, states its aim in a bold subtitle: *Finding Nature's Deep Design*. The question at hand is the same one that Plato asked more than two thousand years ago. Does the world embody beautiful ideas? For Plato, the word *idea* was interchangeable with *form* (and anyone who considers themselves an idealist can trace their aspirations back to ancient Greece). On the mathematical side, Wilczek points to Pythagoras, who shared the same dream that nature would turn out to conform to a perfect geometry.

This belief died hard, but it did die, so why would two acclaimed physicists resurrect it? In Wilczek's version, quantum physics has already exposed a "deep reality" that he terms the Core. There is enough hard evidence to suggest that all the laws of nature and principles of physics are unified at the Core. The classical ideal of planets traveling in perfect circles hasn't survived, Wilczek says, but in the quantum era, "[t]he most daring hopes of Pythagoras and Plato to find conceptual purity, order, and harmony at the heart of creation have been far exceeded by creation." You may think that this is the harmony of an advanced mathematician, and too abstract to translate into beauty in the material world; that we would be left with the same yawning gap between quantum reality and everyday reality. It was this gap that motivated physicists to look for an underlying design in the first place.

Wilczek is capable of waxing eloquent in terms that anyone can appreciate: "There really is a Music of the Spheres embodied in atoms and the modern Void, not unrelated to music in the ordinary sense." *Harmonia mundi*, the "music of the spheres," was a cherished goal for many classical astronomers, including Johannes Kepler. When he made his famous discovery of planetary motion, Kepler considered it a secondary achievement on the way to proving the existence of *harmonia mundi* (a discovery that would mean the angels really do sing).

Note the push-and-pull motion as Penrose and Wilczek fit

the human world into their theories. Penrose is openly distrustful of how the individual mind operates, restating the old, conventional mistrust of subjectivity. That's why he wants to give mathematical structures a reality of their own: "For our individual minds are notoriously imprecise, unreliable, and inconsistent in their judgements. The precision, reliability, and consistency that are required by scientific theories demand something beyond any one of our individual (untrustworthy) minds."

Wilczek is more humanistic; he reveres beauty and wants to rescue the ancient ideal of man as the measure of all things. One of the key illustrations in his book is the famous drawing by Leonardo of a naked man with his arms and legs pictured in two positions. In the first position the limbs fit inside a perfect circle; in the second they fit inside a square. There's an ancient mathematical riddle referred to here, known as squaring the circle. Centuries ago geometry could use simple instruments like calipers and rulers to match up squares, triangles, and other straight-line forms. They also hoped to do the same with the circle. The challenge was to take a circle of a known area and construct a square with the same area in a finite number of steps.

The problem was never solved, but Leonardo's drawing is like a clue pointing in the direction of the human body. Wilczek is very sympathetic to this kind of thinking: "His drawing suggests that there are fundamental connections between geometry and 'ideal' human proportions." This idea goes back to an even more ancient conviction that the universe is mirrored in the human body, and vice versa. "Sadly, perhaps, we humans and our bodies don't figure prominently in the world-picture that emerges from scientific investigation."

Because they consider themselves realists, the vast majority of practicing scientists will regard the word *ideal* with the same distrust that they do *design*. Wilczek and Penrose find themselves facing a steep uphill climb. You'll remember the anthropic principle, which attempts to restore human beings to a privileged place in the universe. Penrose's eternal mathematics doesn't square with that, and Wilczek details a number of objections (as we have, too) that make anthropic thinking iffy, but iffy or not, many roads diverge as soon as anyone tries to connect humans and the cosmos by design. Of course we're attached to the universe as our home, but saying that this connection is part of the cosmic blueprint hasn't led to any sort of settled agreement.

Will it ever? Earth's biosphere is an island of negative entropy that has no scientific reason to exist except that it does exist. The same may be true of cosmic design. Physics may never write the magic equation that brings form out of chaos, but nature is filled with pattern, structure, and form anyway. In broad terms, modern physics is content to believe that the Core, or deep reality, is subject to orderly unified principles. With a bit of hedging, most scientists also accept that mathematics transcends life on Earth and the fallible human mind. Numbers are a truth waiting to be found, but their existence won't change whether anyone finds them or not.

Clearly these two points of agreement aren't enough to build the human universe on, not by themselves. The remaining mysteries are about filling in the gap. It won't do to act as if human beings are incidental specks in a cold, empty void where randomness rules absolutely. No matter how many physicists persist in this viewpoint, there's no denying that humans are woven into the very fabric of creation. Just how far this goes will determine if we are co-creators of a cosmos that begins with the human mind, not the big bang. There may be no other alternative that fits the facts, and fitting the facts is what science is all about.

IS THE QUANTUM WORLD LINKED TO EVERYDAY LIFE?

History has produced more than its fair share of monsters, and when we think about them, we wonder how they could live with themselves. Not just millions but tens of millions of people perished as a result of the actions of Hitler, Stalin, and Chairman Mao. It's chilling to see the home movies of Hitler playing with little children, taking time off from being a monster to assume the part of a smiling uncle.

Why didn't guilt set in? One explanation traces back to an aspect of human psychology that's fairly common, called splitting, also known as black-and-white thinking. Splitting occurs when a person can't join the positive and negative sides of their personality. We all compartmentalize our psychology, keeping under wraps what we don't want other people to see, but splitting takes it to the extreme, allowing someone to be a monster and a nice person without the two sides ever meeting. When neighbors of serial killers invariably describe the murderer as normal and nice, this may be evidence of splitting. The price of living with monstrous deeds is to separate existence into two compartments that do not communicate.

Splitting has a scientific side, too, if we use the term metaphorically. As we've touched upon several times, Einstein's relativistic model is extremely accurate at describing how gravity works and the behavior of large objects in space-time, while quantum theory is equally precise in its description of how the other three fundamental forces work and the behavior of very small objects. The importance of this schism seems abstract. If you know how everything, large and small, behaves, isn't that the same as complete knowledge?

The problem comes down to a simple fact that affects all of us. There is only one reality, not two. A person who has split off his monstrous side is still responsible for what the split-off part did. In court, the good side isn't let off while the bad side goes to jail. Physics has lived with its split for over a century, attempting to unify reality but with limited success. This is a case in which ordinary people have a stake, because the way we live our lives depends on what we accept as real. It was inconceivable in the Middle Ages to live one's life while shutting out God. In an age of faith, nothing was more real than God, and excluding its reality, as it seemed, would be tantamount to delusion, a crime against nature, and surely lead to eternal damnation.

Today we live our lives blithely without paying any attention to the quantum world, and no one is being accused of delusion or heresy; it seems innocuous to split off this most fundamental level of reality. But in this book we contend that reality is basically human, and that claim doesn't hold water if the quantum world is excluded. It's precisely quantum behavior that matters the most. Here's a prime example. In the game of Scrabble you look at your letter tiles, and you have A, O, R, S, S, S, U, U, which looks pretty hopeless. Then you notice that another player has put the word ALL on the board. With a triumphant cry, and a slightly pitying smile, you can now use all of your tiles to make ALLOSAURUS and earn a whopping bonus.

At first glance this little victory has nothing to do with the split between relativity and quantum mechanics, but in fact you have been inhabiting both worlds the whole time you played

Scrabble. Shuffling letters around to make words is a "large object" activity. You have to assemble the right pieces to make sense of scrambled letters. Yet your brain doesn't go through this procedure when you are choosing which word to speak. Mentally you pluck the word you want to say, and the brain delivers it; there is no searching among alphabet letters. For every word in your vocabulary, the spelling, meaning, and sound are fused into one concept, not assembled from scattered parts.

In general, your brain forms connections between billions of neurons, often in widely separated regions of the brain. What's mysterious is how these connections can operate, instantaneously and without any visible communication. The processing speed of neurons can be measured, but that's a different issue from how scattered clumps of neurons "know" to join an activity where teamwork is demanded, as opposed to sending a specific signal down a series of neurons connected like a telephone line. The various patterns needed to coordinate movement, speech, and decision making pop into place automatically. Thus when you see your mother's face, it comes to mind as a face you recognize, not random noses, eyes, and ears that must be examined individually. This looks like something akin to quantum behavior, if nothing else, because cause and effect don't proceed one step at a time. If your mind had to proceed in a linear fashion, step by step, recognizing your mother's face would be like the following:

> Caller 1: Hi, cerebral cortex, this is the visual cortex. You left a message?
> Caller 2: Yeah, I want to see my mother's face. Can you help?
> Caller 1: Certainly, hold on. Okay, I've retrieved some likely eyes. Let's start with those, because most people remember their mother's eyes pretty vividly. We'll get to the other parts once you pick the right eyes.

CALLER 2: Okay. Look, I'm on a schedule. How long is this going to take?

The dialogue sounds comical when slowed down, but even if all the separate parts of your mother's face were being assembled at lightning speed, it wouldn't be instantaneous and holistic. Yet the brain produces the three-dimensional world instantaneously and holistically, exactly the way the quantum world produces large objects like mountains, trees, and everyone's mother.

Leaving the quantum world out of your lifestyle is the same as leaving out your brain. No one does this in reality, of course, because the brain is an absolute essential every minute of our lives. What we do exclude is the connection to the quantum world. This has cosmic implications. For decades a quip has circulated that's attributed to Sir Arthur Eddington: "Not only is the universe stranger than we imagine, it is stranger than we can imagine." It turns out that the attribution was false—the actual speaker remains unknown—and the insight may be wrong, too. The universe may precisely fit what we can imagine. Instead of a universe where particles, atoms, and molecules are mind-like, it seems more likely to us that the universal mind has a way of displaying and acting matter-like. The issue can't be settled until a new mystery is confronted: Is the quantum world linked to everyday life?

GRASPING THE MYSTERY

There is no doubt that quanta are part of the everyday world. When plants convert sunlight into chemical energy, a quantum, the photon, is being processed. Quantum activity is also thought to enable birds to navigate on long migrations by following the earth's magnetic field. Processing electromagnetism in the bird's nervous system would be a quantum effect. Even so, the division between quantum behavior and the ordinary things we experience is crucial in physics. A specific name, the Heisenberg cut, was given to the dividing line that separates quantum events from our perception. Heisenberg himself didn't propose the name—it was bestowed later in his honor—but his thinking repeatedly indicated that there was a (theoretical) line dividing how quantum systems behaved in their own right—as waves—and how they behaved when observed by human observers. He was speaking mathematically. The wave function is one of the chief features of quantum mechanics, but as we've pointed out several times, this elegant construct has never actually been seen in nature. It has to be inferred.

The Heisenberg cut is useful, not so much to divide the real world but to divide the kind of mathematics that works on one side of the line or the other. It's like a border where only French is spoken on one side and only English on the other. But this begs the question of whether quantum reality really is isolated and separate from everyday reality. Perhaps quanta are making things happen all around us that we don't notice. Or maybe the whole picture has been turned upside down—quantum behavior could be the norm in the everyday world, and we only happened to discover it first in the microscopic world of waves and particles.

Not every theory of the universe requires the Heisenberg cut (the multiverse doesn't, for example), but without a doubt the quantum lies at the horizon of our senses. We cannot visualize quanta, and now that dark matter and energy must be confronted, we may have reached the limits of what we can think about. What lies beyond the horizon is both everything and nothing. It is everything because the virtual quantum domain contains the potential for every event that has occurred or ever will. It is nothing because matter, energy, time, space, and we ourselves originate somewhere that's inconceivable. It becomes quite mysterious to reconcile the duality of everything/nothing in order to describe how creation operates.

LIGHT BEHAVES STRANGELY

To get a better idea of the implications for daily life, we'll examine the single experiment that lies at the heart of quantum mechanics, the double-slit experiment, which has a history going back as far as 1801. Early experimenters were interested to see if light waves behave the same as waves of water, for example.

If you drop a pebble into a still pond, its impact will send out rings of waves in circles. If you drop two pebbles into the water a foot apart, each will form a set of rings, and where the two meet, an interference pattern is formed that's separate from the overlapping rings. In quantum physics, this basic fact about wave interference embodies an enigma. In the classic double-slit experiment, a focused stream of photons (light particles) is broadcast onto a screen that has two slits cut into it. The photons that pass through the slits are then detected on another screen placed behind the first (a photographic plate would serve as a simple light-detecting screen). Each photon can supposedly pass through only one slit, and when it gets detected, what appears is a point, the way a pea shot through a pea shooter would leave a pinpoint where it hits.

But if you shoot many photons through the double slits, where they land on the detection plate forms a bar pattern typical of the interference pattern made by waves. In the everyday world, this wouldn't seem possible. It's as if a crowd of people walked through two separate doors to get into an auditorium, and after they sat down, it was discovered that every other seat was filled by a Democrat and a Republican, even though the people entered without giving any political affiliation. Photons passing through a slit individually have no prior affiliation with other photons, yet they gather on the other side in the pattern of a wave, not randomly like the scatter shot of pellets hitting the screen. It is as if each individual quantum, going one at a time, interferes with the other quanta, even if they come in "later."

The double-slit experiment is the classic validation of the particle-wave duality of quanta. So the big question is why two opposite behaviors coexist. In physics we say they are *complementary*, which is more accurate than *opposite*, because the same photon can display either behavior. Keep this "complementarity" in mind, because it holds enormous possibilities. In a universe where A no longer causes B, it turns out that A and B can be two sides of the same coin. To give an example from the natural world, in Africa lions and gazelles share the same watering holes. In the scheme of things, lions eat gazelles, and gazelles run away from lions. But when it comes to water, they coexist. The lions can't keep the gazelles from drinking entirely or else their prey would die of dehydration. The gazelles can't run away automatically, because then they would get no water. Over millions of years, the two species have found a way to make complementary compromises with their opposite roles as eater and eaten.

Over time, the double-slit experiment became more complicated and intriguing. Quantum physics, as we've seen, depends for its lifeblood on measurement and observation. More than in any previous science, how an observer affects the measurement he's making enters into the equation, to the extent that von Neumann believed that quantum reality itself must have a psychological component. Is the observer changing the outcome of the double-slit experiment? The two sides of complementarity, wave and particle, can't be observed at the same time. (In terms of experimental technique, it has also been enormously difficult to observe photons in the first place because they are absorbed by the detector the instant they make contact. But the double-slit experiment is known to work with other particles, like electrons, and has even been roughly duplicated using molecules as heavy as those containing 81 atoms.)

HOW DO PHOTONS MAKE DECISIONS?

It makes physicalists very uncomfortable when the talk turns to photons making decisions and choices or changing their properties depending on how they are being observed. Starting in the late 1970s, John Archibald Wheeler developed a series of thought experiments to test the crucial question. Do photons change their behavior because of the questions/intentions of the experimenter? The alternative is that they change their behavior for some purely physical reason, such as interacting with the detector device.

Wheeler's thought experiment considered how a photon actually behaves in flight. Remember, a photon can't be seen in flight and is known only at the moment of detection. If a detector is placed right at the slit, it shows in real time that each photon passes through one slit, the way a pellet would. What if we place the detector after the slit? Wheeler asked. It turns out that the photon can delay its decision to behave like a wave or a particle until after it passes through the slit, which is very peculiar. But it was just as peculiar to suppose, as some theorists did, that in wave mode, a photon would be passing through both slits at the same time.

Going a step further, could photons make decisions and later change their mind? This is a distinct possibility in Wheeler's thought experiment. For example, you can place two aligned polarizers at the double slits to cancel out any wave-like interference, yet if you then let the photons pass through a third polarizer that erases this effect, the photons will be restored to their original state and can behave like waves, producing the interference pattern that supposedly was erased.

This twin phenomenon of "delayed choice" and "quantum eraser" makes it hard to believe in a strict physicalist explanation; the way a quantum is observed takes center stage. There were other wrinkles, too. Physicist Richard Feynman proposed that if a detector for individual photons were placed between the two slits, the wave-like interference pattern would disappear. Both Wheeler's and Feynman's thought experiments have been generally accepted, despite great difficulties in setting up actual lab experiments to validate them. But do they solve the mystery of what the observer is doing to make photons behave the way they do? Like a ghostly apparition, the observer effect appears before our eyes, but we can't wrap our arms around it.

Wheeler hit upon the right conclusion, we feel. He declared that physicists were making a mistake to believe that particles had the dual properties of wave and particle to begin with. "Actually, quantum phenomena are neither waves nor particles but are intrinsically undefined until the moment they are measured. In a sense, the British philosopher Bishop Berkeley was right when he asserted two centuries ago 'To be is to be perceived.'"

In other words, there is no observer "effect" or "problem," as if the observer is an intruder who pops in on nature, disturbing her privacy by peeking this way and that. Instead, things exist because they are perceived. This insight on Wheeler's part is why he insisted over and over that we live in a participatory universe. The observer is woven into the very fabric of reality. Suddenly, the human universe doesn't seem either far-fetched or far away.

The quantum revolution is over a century old. Why hasn't the mind-like behavior of the universe become commonly known; why is it not being taught in schools? If anything, the cosmos is more elusive now than it was in the first twenty-five to thirty years of the quantum era. In large part the bafflement being felt today comes back to the Heisenberg cut. A strict division between quantum and classical worlds may work mathematically, but in reality the dividing line is porous, blurry, and perhaps a mirage. If it takes an observer, seated squarely in the classical world, to prompt a photon to make a choice, seated squarely in the quantum world, how alien could the two realms be?

So let's shift the emphasis and ask why we don't perceive quantum effects in daily life. Quanta are very small, but so are viruses, and they exert enormous effects all the time by causing disease. A cold or flu virus comes and goes in your body, but quanta affect you at every moment. Raise your hand and take a look at it. In this simple gesture you've performed a quantum activity, since sight begins with photons, which are quanta, striking the retina of your eye. Look at your garden and the trees outside—photons of sunlight make them grow. So being microscopic isn't a problem with photons. Instead, we have some built-in mechanisms that act as blocks to truly perceiving what photons do.

CAN THE BRAIN BE TRUSTED?

Nothing is real to us until we perceive it, and as it happens, the human brain is a very selective mechanism for perception. It can be as delicate as the most sophisticated photon detector—in essence, that's what the visual cortex is—while at the same time the brain has no knowledge of how its own processes work. You don't possess inner vision that shows you the firing of neurons in your brain. A loud noise makes you jump because an automatic brain mechanism causes your response, but you can't witness it or the stress hormones, like adrenaline, that fuel the fight-or-flight response. The brain's blindness to its own activity is the main reason so many phases of life, like puberty or the effects of aging, surprise us when they arrive.

A major drawback of naïve realism is the assumption that the human brain delivers a picture of reality, when in fact it doesn't. It delivers a convincing three-dimensional image of the world that is nothing more than a perception. Think about the double-slit experiment we just discussed. Most of its difficulty comes from the fact that photons are invisible when in flight and are only detected as they perish. If light is invisible to begin with, there can be no way to make it visible except through a nervous system, and once that is accomplished, light is no longer its natural self but a neural creation.

Change the nervous system and light will change with it. An owl's acute night vision, the ability of an eagle to spot a mouse from hundreds of feet in the air, the underwater sight of dolphins, and a bat's ability to "see" using echolocation—all of these examples are radically different from human sight. Therefore, the assumption that we see "real" light is unfounded. There is nothing about photons that necessarily makes them visible. Billions of stars and galaxies are totally invisible until a nervous system turns them luminous.

Perception is fallible because no two people see the world exactly alike—that's a given. But in many ways the brain's relationship to reality is murky. Alfred Korzybski, a pioneering mathematician, set out to calculate precisely what the brain does when it processes raw data. First of all, the brain doesn't absorb everything but erects a complex set of filters. Some of these filters are physiological; that is, the biochemical apparatus of the brain can't cope with all the signals being conveyed to it.

Billions of bits of data bombard our sense organs every day, of which only a small fraction makes it past the brain's filtering mechanism. When people say, "You're not hearing me" or, "You only see what you want to see," they are expressing a truth that Korzybski tried to quantify mathematically.

But other filters are psychological—we don't see and hear certain things because we don't want to. Perception can be distorted by stress and high emotion, or any number of mixed signals in the brain. For example, if you are alone in your house at night and hear a loud creaking sound, you will react with alarmed alertness, because your lower brain, which is responsible for basic survival, has a privileged pathway when it detects possible threats. It takes a moment or two before the higher brain, the cerebral cortex, gets your attention. It decides whether the creak was a possible intruder or just a noise in the rafters or the floorboards. Once you make a rational determination, your brain mechanism can allow a balanced response, based on a clear assessment of the situation.

Fire up the lower brain's survival mechanism too much, which is what happens to soldiers at the front line under constant artillery bombardment, and the brain is prevented from returning to a state of balance. The inevitable result, no matter how brave and stalwart a soldier may be, is battle fatigue or shellshock. When the brain's coping ability has been overstressed, its perceptions turn completely unreliable.

Then again, sometimes the limitation isn't about filtering. The things a person can't perceive may simply lie outside the range of what human sense organs can perceive, like our inability to see ultraviolet light or to hear ultrasound. Still, a great deal of distorted reality depends on expectations, memories, biases, fears, and willfulness. "Don't bother me with the facts, my mind is already closed" is too true to be very funny. Instead of filters, we are dealing with self-created censors, mental watchdogs that shut out certain information because it is personally unacceptable. Who would date a man who was the spitting image of Hitler or Stalin? If you go to a party, and someone tells you that you are about to meet a Hollywood star, you will see a different person than if you had been told he is a convict on parole. Taking all of these selective limitations into account, they make it very clear, as Korzybski pointed out, that the brain is extremely fallible when it reports reality.

But that's only the beginning. The brain can be trained, and everyone's brain has been. It accepts only the model of reality it was trained to accept. That's why the worldview of a religious fundamentalist isn't shaken by scientific facts—they simply don't compute according to the model his brain accepts. The model of reality you are following right this minute is wired into the synapses and neural pathways of your brain. Consider a shabbily dressed old man walking down the street. Passersby see the same visual information, but to some the old man is invisible; to others he's an object of sympathy; to others still he's a social menace or dead weight or a reminder to call their grandparents. He is the same man but produces a vast number of perceptions among a vast number of perceivers. Even to the same perceiver, it's unavoidable that perception changes with time, mood, recollection, and so on.

We may assume that we are in control of our responses to the world, but that's far from the case. If two people can see the same thing and have opposite reactions, their responses are controlling them, not the other way around.

Science prides itself on following a rational model, yet even so, there are certain undeniable facts that undermine rationality. Every brain has been trained to perceive the world in ways we can't escape, no matter how rational we believe we are. If you were told that a thousand strangers you've never met will die unless you commit suicide, rationality would be a poor motivator: your brain is programmed for survival. At the same time, soldiers will sacrifice themselves in battle to save a buddy, because courageous altruism is part of a soldier's code, overriding the survival instinct.

Models are powerful things. Yet it's important to realize that reality transcends all models. John von Neumann is credited with saying that the only satisfactory model of a neuron would be a neuron. In other words, models are no substitute for the complexity and richness of what occurs naturally. Or, as Korzybski put it, "The map is not the territory." Even the best map of a city, if it gave you three-dimensional real-time moving pictures from a super-GPS, could never be mistaken for the real city.

Every model has the same fatal flaw: it throws away the things that don't fit. Subjectivity doesn't fit the scientific method, and so the vast majority of scientists throw it out. Physicalists throw out mind as a force in nature. Because of this inherent flaw, models are right about what they include and wrong about what they exclude. In our view, the last person to ask about the mind is a physicalist, just as the last person to consult about God is an atheist.

We are forced to a startling conclusion: no one can claim to know what is "really" real as long as the brain is their window on the universe. You can't step outside your nervous system. Your brain can't step outside space-time. So whatever is outside time and space is a priori inconceivable. Unfiltered reality would probably blow the brain's circuits, or simply be blanked out.

All of these facts seem to prove that we live on the classical side of the Heisenberg cut. That's a false conclusion, however. Everything we say, think, and do is connected to the quantum world. Because we're embedded in quantum reality, we must be communicating with it somehow. The quantum state is just as available as the everyday world. Going into the quantum state doesn't mean that every solid object becomes illusory and all your friends are imaginary. It means that you've stepped into another perspective, and by perceiving your life as a multidimensional series of quantum events, that's what it becomes.

ADAPTING TO THE QUANTUM

You have a quantum mechanical body, including your brain, which means that the self you call "me" is a quantum creation. The world is no different. Quantum theory is the best guide, so far, to how nature actually operates. Even though strict believers in the Heisenberg cut don't permit the classical and quantum worlds to bleed into each other, clearly they do. Does this mean that you behave like a photon, and vice versa? Yes. A prime example is unpredictability. In classical physics, the whole point was to tame nature's messiness, making events "out there" abide by rules, constants, and laws of nature. This project was spectacularly effective until quantum mechanics became the new sheriff in town.

At that point, unpredictability became a fact of life, just as it is in human behavior.

Each unstable radioactive nucleus has a specific rate of decay known as its half-life, the amount of time that it takes for it to lose half of its initial value. The half-life of uranium 238 is about 4.5 billion years. In general radioactive decay is very slow, which is why sites contaminated by radiation can be dangerous far beyond our lifetime. The process is also unpredictable, in that a physicist can't point to a specific nucleus and tell us when it will decay. Therefore probabilities are given instead—this is a key adaptation to quantum reality. Uncertainty is a given.

For the sake of illustration, if a certain nucleus has a half-life of a day, it will have a 50 percent chance of decaying within a day, a 75 percent chance within two days, and so on. The quantum mechanical equation (specifically the Schrödinger equation) that describes a particular quantum system is very precise about the probability of an occurrence in the nucleus. But a problem arises. It is an obvious fact that any probability refers to something that's about to happen, whether it concerns the outcome of nuclear decay or the winner of the Kentucky Derby. But after it happens, the outcome suddenly jumps to 100 percent (decay occurred, American Pharoah won the Derby) or else 0 percent (no decay occurred, another horse won). The probabilities of real-life events must at some point jump to 0 or 100 percent once the outcome is known. Otherwise they mean nothing.

The Schrödinger equation calculates the "survival probability" of a nucleus (i.e., the probability of its not having decayed), which starts off at 100 percent, and then falls continuously, reaching 50 percent after one half-life, 25 percent after two half-lives, and so on—but never reaching zero. (Good news for slow racehorses, who will get infinitesimally closer to the finish line but never cross it to be declared the loser.)

So, as spectacularly successful and honored as the Schrödinger equation has been, it never describes *an actual event*! If there was an actual decay, the survival probability would become a certainty and jump to 100 percent at that point, because we are certain that the decay occurred once we observed it. This gap between mathematics and reality has become famous as the paradox of Schrödinger's Cat, a thought experiment devised by the great man in 1935, which has defied explanation ever since, even though every theoretical physicist has their own pet answer.

A PARADOXICAL CAT

The experimental setup is that Schrödinger has put his cat inside a steel box and closed the lid. Besides the cat, the box also contains a tiny lump of radioactive material, a Geiger counter, and a flask of poison. The lump of radioactive matter is small enough that one of its atoms may or may not decay in the space of an hour. The odds, Schrödinger proposes, are 50/50. Now, if an atom does decay, the Geiger counter will detect it, triggering a trip-hammer that will fall and shatter the flask of poison, killing the unfortunate cat. If no decay occurs, the cat is out of danger, and when the lid of the box is opened, the animal will be alive. So far, these two outcomes conform to common sense.

But not in quantum terms. The two possible outcomes, decay of radioactive material and nondecay of radioactive material, both exist in a superposition (a bleary state). According to the Copenhagen interpretation, which prevailed at the time, it takes an observer to cause a superposition to collapse into a specific state. No one could quite explain how the observer actually did this, but until an observer comes along, quanta remain in superposition, treading water as it were.

If your head swims thinking about this famous thought experiment, it's reassuring to know that Schrödinger himself found superposition to be absurd when it came to real life. If the nuclear decay of the radioactive substance is in superposition, he argued, then according to the Copenhagen interpretation, before the box is opened, its state is suspended 50/50 until an observer appears. Which may be good enough for a quantum, Schrödinger argued, but what about the cat? It would be dead and alive at the same time, suspended 50/50 between the two states, until an observer opens the box! It's alive insofar as the atom didn't decay; it's dead insofar as the atom did decay and released the poison.

Of course, a cat can't really be dead and alive at the same time. Everyone agreed that this was a most clever paradox, but it takes a bit of close thinking to grasp why. Schrödinger's Cat is all about the gap between quantum behavior and real life. The "smeary" state of superposition makes no sense in the real world, where a cat is either dead or alive, not waiting for someone to look at it before its fate is decided.

Einstein was delighted by this thought experiment and wrote to Schrödinger:

> *You are the only contemporary physicist . . . who sees that one cannot get around the assumption of reality, if only one is honest. Most of them simply do not see what sort of risky game they are playing with reality. . . . Nobody really doubts that the presence or absence of the cat is something independent of the act of observation.*

Unfortunately, the paradox is not as simple as Einstein would have it. In the so-called many-worlds theory proposed by physicist Hugh Everett, the cat is dead and alive at the same time, but in different realities or worlds. Quantum outcomes are not either/or but both/and, depending on which world you stand in. When the box is opened, Everett's explanation goes, the observer doesn't magically cause an outcome; rather, there is both an observer seeing a dead cat and an observer seeing a live cat. These two equally real scenarios split off from each other with no communication between them. The one observer won't be aware of the other.

Like the multiverse, the many-worlds theory is nifty in the way it turns head-scratching problems into no problem at all. You can have your cat and kill it, too. But exactly how these split realities shear off (known as quantum decoherence) injects a new problem, and since other worlds are just as theoretical as other universes, it's hard to believe that they aren't imaginary, purely mathematical fancies. The net effect of the many-worlds interpretation is that the challenges created by the Copenhagen interpretation are magnified to infinity!

Perhaps Schrödinger's Cat is trying to tell us something completely different. Instead of seeing quantum behavior as exotic, paradoxical, and far removed from ordinary life, it could be that all of us are existing in a quantum state already, and quanta are just imitating us. If we ask whether Schrödinger's Cat is dead or alive inside the box, the possible answers are yes, no, both, and neither. Why does this appear so paradoxical? If a boy takes a girl to see the latest Marvel Comics movie and asks her if she wants popcorn or a Coke, she might say yes or no to either one, pick both, or want nothing at all. This is naturally how free will works. Choice is open to all possibilities until a choice is made.

Let's put the girl in Schrödinger's box, without the poison and the radioactivity. Before we open the box to find out if she wants popcorn or a Coke, in what state is her answer? Is it a superposition of yes, no, both, or nothing? The answer is that this is the wrong question to ask if you know how the mind works. The girl is simply waiting to make up her mind. Her answer doesn't inhabit an exotic limbo, like an atom smeared between decaying and not decaying, but the two situations aren't completely different. Even though we have thoughts all the time, we don't know where they exist before we think them. By the same token we don't know where our next word exists before we speak it.

In fact, being able to call up a word out of thin air is rather miraculous. If you want to tell a friend that you saw the pandas at the Washington Zoo, you simply say so. You don't rifle through a mental library of Chinese mammals until the right verbal tag is found. A computer can't duplicate this everyday feat. It must consult a storage bank of programmed memories in order to match word and meaning. (In fact, no computer knows the meaning of any word.)

You could say that thoughts and words are in a kind of silent limbo, waiting to be called upon by the mind. Words are just possibilities waiting to emerge into the world, as quanta do. Wheeler touched on an important point about reality when he said that quanta don't have properties until they are perceived. Likewise the contents of our minds. Try to describe what your exact thought will be at noon tomorrow. Will it be angry, sad, happy, anxious, optimistic? Will you be thinking about lunch, work, family, or the weekend football game?

You can't make an accurate prediction because a thought, like a quantum, has no properties before it pops into existence. There's no mystery to this if we honor Einstein's warning that we shouldn't play games with reality. What physicists called quantum indeterminacy stands for the fact that quanta can't be known until the very moment of measurement. The same is true of thoughts, words, human behavior, and the evening news. The reason we rush to find out the latest disaster on the evening news is that we are well adapted to reality as a messy, unpredictable thing, rough around the edges, and ruled by uncertainty. The quantum revolution didn't introduce these elements into our lives; it merely expanded them from the human into the quantum world.

Are we ready now to take the big leap and say that human beings created the quantum world? Not quite. The issue of how an observer affects reality hasn't been settled. Some very strange quantum behavior still needs to be domesticated. But we've reached a turning point. The Heisenberg cut, in real-life terms, is a mirage. All of us live in the multidimensional quantum world. We project ourselves into everything we experience, not just by observing but by participating in the reality that emerges. When we do this, are we being self-centered, injecting human qualities into the universe because it suits our vanity? Or did the universe already contain mind in the first place? That's the hot-button issue lying at the heart of the next mystery.

DO WE LIVE IN A CONSCIOUS UNIVERSE?

For the average person, the notion of infinite universes bubbling up here, there, and everywhere is a nice piece of imagination, or could be pondered as weird science. In any case, there are many skeptics to challenge the multiverse, and as the argument rages, a bystander might raise his hand and ask, "Do we actually know what *this* universe is like? Never mind all those others."

It's a valid point. The multiverse is like a romance novel for the whole human race. In romance novels, the heroine ultimately finds Mr. Right. In the multiverse, human beings have found Mr. Right Cosmos. (Except that the odds of finding the right cosmos are essentially zero, vanishingly smaller than the chances of finding Mr. Right in everyday life.) The only question is whether, like our Harlequin heroine, fate made the perfect match or it was simply dumb luck. In this book we're saying it is neither. The perfect match between human beings and the universe is about a meeting of the minds. The human mind matches the cosmic mind. In some mysterious way that science hasn't explained, we find ourselves living in a conscious universe. Or to be truly mind-bending, we live in an unbounded state of consciousness that we call the universe.

Massive skepticism would greet this proposition at a typical physics or neuroscience conference, but we've already seen the mounting evidence of how the quantum domain acts in a mind-like way. This evidence has been studiously ignored. In modern physics, consciousness has been like a black hole, swallowing up every investigator who has tried to give definitive answers. No one has ever written a book titled *Mind for Dummies*, because the topic has defeated, and continues to defeat, the most brilliant thinkers. Humans are in the ironic position of knowing for certain that we have a mind, while at the same time finding that our mind can't explain itself. Just asking "Where does a thought come from?" leads to bafflement, loud arguments, and a severe headache. Yet the beauty of a conscious universe lies in how many questions it solves at one stroke, as follows:

Q: Are human beings the only conscious creatures on Earth?

A: No. All living creatures participate in cosmic consciousness. In fact, all so-called inert objects participate in it, too.

Q: Does the brain produce the mind?

A: No. The brain is a physical instrument for processing mental events. Mind and brain can both be traced to the same source: cosmic consciousness.

Q: Is there consciousness "out there" in the universe?

A: Yes and no. Yes, there is consciousness everywhere in the universe. No, it isn't "out there," because "in here" and "out there" are no longer relevant concepts.

The simplicity of these answers is what appeals to any scientist who accepts the possibility of a cosmic mind. We are steadily climbing out of the black hole. Today there are papers, books, and conferences devoted to the conscious universe, and a mini-revolution is under way. To be realistic, though, mainstream science still prefers to ignore consciousness.

Science is in the habit of excluding assumptions that aren't necessary to solving a problem. In the working world of physics, it's irrelevant to $E = mc^2$ or the Schrödinger equation or chaotic inflation if the universe is conscious. A huge amount of productive science has emerged by excluding the entire issue of the mind. (Just as treating a baby as a puppet is workable at a certain level.)

But that's not the really peculiar part. What we find eminently strange is that scientists consider their own minds irrelevant. It's simply a given, like breathing. When someone is bombarding protons in a particle accelerator, nobody says, "Make sure you're breathing," much less "Make sure you're conscious." Both are irrelevant assumptions. And yet, looked at another way, nothing is more important than the mind, especially if the human mind is somehow in sync with a cosmic mind. It matters to all of us if human beings have a cosmic dimension. All talk about being merely a speck in the vast coldness of outer space would come to an end forever. As Wheeler poetically put it, we are "the carriers of the central jewel, the flashing purpose that lights up the whole dark universe."

GRASPING THE MYSTERY

The main roadblock to a cosmic mind is the assumption that the mind is always tainted by its subjectivity. Subjectivity is alien to data and numbers, the stuff that makes science a viable activity. General agreement is reached by studying the facts and nothing but the facts. In consciousness studies, however, objectivity is classified as a separate variety of human awareness, known as third-party consciousness, meaning that any third party can come on the scene and agree with what has been observed. For example, consider a team of geologists picking over the ground at Point Trinity, the spot in the New Mexico desert where the first atomic bomb was exploded on July 16, 1945. The first geologist spies an unusual mineral lying on the ground. As they examine it, the second geologist agrees that it looks like nothing he's ever seen before.

The rock specimen is tested by other geologists, and a consensus is reached. The enormous heat of that first atomic blast created a mineral unknown anywhere else on Earth, which they named trinitite. The desert sand, composed primarily of quartz and feldspar, was fused into this glassy green residue, which is mildly radioactive but not dangerous.

The discovery of trinitite neatly conforms to third-party consciousness. By eliminating all subjective reactions (known as first-party consciousness), objectivity is assured, or so they say. There is also second-party consciousness, the "you" that sits across the table from "me." Second-party consciousness is almost as untrustworthy as first-party consciousness, since two people can share the same delusion. No one has shown how to go from two observers sharing the same experience to actual objectivity.

Throwing out any reference to consciousness except for the third-party kind is enormously convenient if you're a physicalist. It also sweeps a huge amount of experience under the carpet, all along saying this is the only way to do science. Looking around at the modern world, which was built on science and technology, one is looking at the vast possibilities of third-party consciousness. You can see why science is so eager to throw out first-person consciousness, the "I" of everyday experience. Rembrandt can say "That's my self-portrait," but Einstein can't say, "That's my relativity. If you want some relativity, get your own."

Yet, by making third-party consciousness the norm, we wind up with a science-fiction world where no "I" exists. To see the weirdness of the situation, try walking around and referring to yourself only in the third person. *He just got out of bed. She is brushing her teeth. They seem reluctant to go to work, but they have to put food on the table.* It cannot be denied that subjectivity is messy, but it's also how experience works. Things happen to people, not to pronouns.

Naturally, every scientist has an "I" and a personal life. But in the models of reality developed by physics and modern science in general, the universe is a third-party experience. As John Archibald Wheeler famously said, it's as if we look at the universe through a foot-thick piece of glass when what we should be doing is breaking the glass.

An unconscious universe is a dead universe, while the universe that human beings experience is alive, creative, and evolving toward magnificent structures that are even more creative. If the latest data from the Kepler observatory are valid, the number of earth-like planets in the observable universe may be as many as 1 followed by 22 zeros. The enormous number of planets that might sustain life could be proof that a conscious universe is expressing itself many times over.

The argument about how humans evolved on Earth can't be settled as long as consciousness itself remains a mystery. When we talk about it, consciousness needs to be clear, reasonable, and believable. No mode—first-person, second-person, or third-person—can be banished. There must be a level playing field, with no pronouns playing favorites just because they can get away with it.

WHEN ATOMS LEARNED TO THINK

Everything in the cosmos is either conscious or unconscious. Or, to be more precise about our terms, an object is either participating in the domain of mind or it isn't. Choosing which is which, however, isn't as easy as it appears. Why do we say the brain is conscious? The brain is made up of ordinary atoms and molecules. Its calcium is the same as the calcium in the White Cliffs of Dover; its iron is the same as the iron in a two-penny nail. As thinkers, nails and the White Cliffs of Dover aren't famous, but we all accept that the human brain has a privileged place in the universe, meaning that its atoms are somehow unique compared with the same atoms in "dead" matter.

When a molecule of glucose passes through the blood-brain barrier (a cellular gatekeeper that determines which molecules are allowed to pass from the bloodstream into the brain), the glucose doesn't change physically. Yet somehow it contributes to the processes we call thinking, feeling, and perceiving. How can the simple sugar regularly used to nourish hospital patients through an IV tube learn how to think? That question goes to the heart of the mystery. If all objects in the universe are either part of consciousness or are not, the conscious ones learned how to think, and yet no one has ever explained how this occurred.

Really, the whole notion of atoms learning to think is totally irrational. The exact moment when atoms acquired consciousness will never be located. Linking mind and matter has been labeled "the hard problem" and has become the focus of intense debate. Out of the 118 elements found, only 6 make up 97 percent of the human body: carbon, hydrogen, oxygen, nitrogen, phosphorus, and sulfur. If anyone hopes to mix and match these atoms in such a hugely complex way that they suddenly start to think, this would seem like a naïve goal. But in essence that's the only explanation offered for how the human brain became the organ of consciousness.

With billions of base pairs contributing to the double helix of human DNA, complexity becomes bewildering enough to serve as a plausible cover for ignorance. Telling which objects are conscious and which are not is very tricky. Calling the entire cosmos conscious is just as plausible as calling it unconscious. The argument can't be settled simply on physical grounds.

The mystery boils down to a clear-cut choice: is the universe made of matter that learned to think, or is the universe made of mind that created matter? We can call this the divide between "matter first" and "mind first." Although "matter first" is the default position of science, the quantum century seriously undermined it.

One popular view tries to rescue the "matter first" position by cleverly turning everything into information. We are surrounded by information on all sides. If you receive an e-mail announcing a sale on smartphones, a piece of new information has come your way. Yet the photons that strike your retina as you read the computer screen also carry information, which gets transformed into faint electrical impulses in the brain that are another kind of information. Nothing is exempt. At bottom, anything a person can say, think, or do can be computerized in the form of digital code using only 1s and 0s.

A model can be developed where the observer is a bundle of information looking out upon a universe that is an even bigger bundle of information. Suddenly, mind and matter find common ground. Some cosmologists consider this a viable alternative to a conscious universe. All it takes, we are told, is to define consciousness purely as information. An articulate proponent of this view is physicist Max Tegmark of MIT. He begins his argument by dividing consciousness into two problems, one easy, the other hard.

PROBLEMS EASY AND HARD

The easy problem (which is hard enough) is to understand how the brain processes information. We've made strides in that direction, Tegmark maintains, considering that computers are now advanced enough to defeat the world's chess champion and translate the most difficult foreign languages. Their ability to process information will one day surpass the human brain's abilities, and then it will be nearly impossible to say which is conscious, the machine or a human being. The hard problem is "Why do we have subjective experience?" No matter how much you know about the hardware of the brain, you haven't really explained how microvolts of electricity and a handful of dancing molecules can deliver a person's awe at seeing the Grand Canyon for the first time or the rush of joy that music produces. In the inner world of thoughts and feelings, data get left behind.

"The hard problem" acquired its official name thanks to philosopher David Chalmers, but it has been around for centuries as "the mind-body problem." Tegmark sees a solution by relying on a scientist's treasured ally, mathematics. To a physicist, he says, a human being is just food whose atoms and molecules have been rearranged in complicated ways. "You are what you eat" is literally true.

How is food rearranged to produce a subjective experience like being in love? Its atoms and molecules, from the perspective of physics, are just an amalgam of quarks and electrons. Tegmark rejects a force beyond the physical universe (i.e., God) butting in. The soul is also out. If you measure what all the particles in your brain are doing, he argues, and these particles perfectly obey the laws of physics, then the action of the soul is zero—it adds nothing to the physical picture.

If the soul is pushing the particles around, even by a small amount, science would be able to measure the exact effect the soul is having. Voilà, the soul becomes just another physical force with properties that can be studied the way we study gravity. Now Tegmark unveils the idea that either solves the hard problem or turns out to be a very clever sleight-of-hand. As a physicist, he says, the activity of particles in the brain is nothing but a mathematical pattern in space-time.

Dealing with "a bunch of numbers" transforms the hard problem. Instead of asking "Why do we have subjective experience?" we can look at the known properties of particles and ask a question based on hard facts, "Why are some particles arranged so that we feel we are having a subjective experience?" This may sound like a movie scene where absentminded Professor Brainiac is scribbling equations on the blackboard to explain why he's attracted to Marilyn Monroe sitting in the front row. But Tegmark's trick of turning the subjective world into a physics problem has obvious appeal within his field.

But it's not hard to be skeptical. Einstein's mind produced wonderful calculations; it's unlikely that wonderful calculations can produce Einstein's mind. But Tegmark argues that they can. The things that exist all around us, he says, possess properties that can't be explained simply by looking at the atoms and molecules they are made of. The H_2O molecule doesn't change as water turns into ice or steam. It simply acquires the properties of ice and steam—so-called emergent properties. "Like solids, liquids, and gases," Tegmark declares, "I think that consciousness too is an emergent phenomenon. If I go to sleep and my consciousness goes away, I'm still made of the same particles. The only thing that changed was how the particles are arranged."

We are using Tegmark here to stand for a whole class of thinkers who believe that math holds the key to explaining the mind. In their view consciousness is no different from any other phenomenon in nature. Numbers can be assigned to information, and information is defined by Tegmark and others as "what particles know about each other." At this point a great deal more complexity must be introduced, but you've gotten the key concepts.

The focus burns brightest on the integrated information theory proposed by Giulio Tononi, a neuroscientist at the University of Wisconsin. To bridge the gap between mind and matter, Tononi and his colleagues devised a "consciousness detector" that can be used medically, for example, to indicate if someone who is completely paralyzed still has consciousness. Such a development is intriguing for brain research in many ways.

But information theorists are hunting for bigger game. They want 1s and 0s, the basic units of digital information, to explain consciousness in the cosmos at large. It's true that particles with positive and negative charges can easily be described with a 1 or a 0, and the same holds true anytime a property in nature has an opposite, the way gravity may be coupled with antigravity. But do numbers really help us to get from lifeless particles to love, hate, beauty, enjoyment—all the things happening "in here"? Highly improbable. Knowing that water acquires the emergent properties of ice doesn't get you to ice sculptures. Something else is obviously at work.

We are told that information is "what particles know about each other," but that's the problem, not the solution. The notion that throwing in more and more information will build a full-fledged human mind is like saying that if you add more cards to the deck, they will suddenly start playing poker. Jacks, queens, and aces all carry information, but that's not the same as knowing what to do with the information, which requires a mind.

LETTING REALITY SPEAK FOR ITSELF

Everyone who has tackled the problem of consciousness feels that they have reality on their side. Yet if you look more closely, no theoretical model can tell us what's real. Radar can tell you when it's raining, but only you can tell that rain is wet—experience is the only judge. It's remarkable that the nuclear inferno inside a star can be reduced to 0s and 1s, but the concepts of zero and one are human. Without us, they wouldn't exist.

In fact, there is no information anywhere in nature without a human being who understands the concept of information. With information theory severely undermined, the most common fallback is to say, "We can wait for a better theory someday. Meanwhile, there's new brain research emerging every day. It will tell us the story eventually." But this kind of certainty is based on a very shaky assumption, that Brain = Mind.

The entire field of neuroscience is based on this assumption.

Undoubtedly there is activity in the brain when a person is alive and conscious, while death brings the cessation of this activity. But imagine a world where all music comes through radios. If the radios break down, the music dies. Yet this event wouldn't prove that radios are the source of music. They transmit it, which is a big difference from their being Mozart or Bach. The same could be true of the brain. It could simply be the transmitting device that brings us our thoughts and feelings. No matter how powerful brain scans ever become, there's no proof that neural activity creates the mind.

The problem with Brain = Mind is twofold. First, there's the assumption that the mind is an epiphenomenon, in other words, a secondary effect. If you light a bonfire, the primary phenomenon is combustion; the secondary phenomenon is the heat that the fire gives off. Heat is an epiphenomenon. In brain research, it's assumed that the physical activity inside neurons is the primary phenomenon; the subjective sense of thinking, feeling, and sensing is secondary. Mind becomes an epiphenomenon. Yet it's fairly obvious that being aware of who you are, where you are, and what the world looks like—everything that comes with mind—is just as likely to be primary. Music came before radios, and this fact isn't undermined by studying how radios work down to their atoms and molecules.

The second problem with Brain = Mind is that we have no way to see nature accurately. It's hard to grasp just how complete our blindness to reality is. The narrator in Christopher Isherwood's *Goodbye to Berlin* is a nameless young man who has arrived in Germany during the rise of Hitler. Instead of showing us how appalled he is, Isherwood wants us to make our own judgments, because only then will we believe in the horror of what the narrator sees. The young man begins his tale by saying,

> *I am a camera with its shutter open, quite passive, recording, not thinking. Recording the man shaving at the window op-*

posite and the woman in the kimono washing her hair. Someday, all this will have to be developed, carefully printed, fixed.

But a camera is exactly what the human brain isn't, or the human mind. We are participants in reality, which makes us totally involved. Quantum physics is famous for bringing the observer into the whole problem of doing science, and equally famous for not solving what the observer's role is.

The practice of science didn't grind to a halt waiting for the solution, and therefore a fallback position has been adopted: leave the observer out. For some physicists, this means "leave the observer out for the time being," while for others, the vast majority, it means "leave the observer out all the time—it's not as if he really matters." But reality begins with "I am," minus the camera. Every person wakes up in the morning to face the world through first-party consciousness. It's an inescapable fact.

With two strikes against it, Brain = Mind should be seriously doubted. Ironically, however, the mind needs the brain and can't do without it, so far as we know. Like the imaginary world where radios are the only way to access music, our world has no access to the mind except through the human brain. In his memoirs, psychiatrist David Viscott reported a life-changing incident that happened to him in a hospital when he was in training. He walked into a patient's room just as the patient died, and in that instant he saw a light leave the body, for all the world like the soul or spirit departing.

The fact that he had seen such a thing—which isn't uncommon among hospice workers—shook Viscott's beliefs to their core. His worldview couldn't account for such a phenomenon, and he knew that his medical colleagues wouldn't believe him. If they had a soul, that didn't mean they believed in souls. Likewise, even if your brain is just a receiving device for the mind, you can still argue that the brain *is* the mind. (Another proof that your belief system is more powerful than reality.)

FOLLOW THE MOVING ARROW

Is there any way to settle the dispute between "mind first" and "matter first"? If our beliefs stand in the way, perhaps reality must speak for itself, so there is no mistaking the results. One avenue stems from many centuries ago, in a paradox first posed in the fifth century BCE by Greek philosopher Zeno. The common term for it is *Zeno's arrow paradox*.

As an arrow flies through the air, Zeno said, we can observe it at any instant in time. When we do, the arrow occupies a specific position. For the instant that it holds any position, the arrow isn't moving. So if time is a series of instants, it follows that the arrow is always motionless. How can an arrow be moving and motionless at the same time? That's the paradox, and it came to life two millennia later in the *quantum Zeno effect*, a term coined by George Sudarshan and Baidyanath Misra of the University of Texas. This time the object being observed isn't an arrow but a quantum state (such as a molecule undergoing a transition) that would ordinarily decay in a finite amount of time.

A quantum state that should decay is frozen by continuous observations. In many, though not all, interpretations of quantum mechanics, the wave-like behavior of a particle "collapses" into a state we can measure and observe thanks to the observer, although how the observer tinkers with this transition is highly controversial. As we've seen, the actual moment when a molecular state would decay can't be determined but only estimated, using probabilities. But in the quantum Zeno effect, the intervention of observation changes the system from an unstable one to a stable one.

Can you stand around watching a molecule constantly, to see when the actual event occurs? No, and that's the paradox. If an observer watches continuously or at superfast intervals, the state being observed will never decay. As with viewing a flying arrow in chopped-up instants of time, observing unstable quantum systems subdivides the chopped-up activity so finely that nothing happens. By analogy, imagine that you are a wedding photographer taking a picture of the bride. When you say "Smile," the bride says, "I can't smile when the camera is pointing at me." Now you're stuck. As long as you have the camera on her, there will be no smile. If you take the camera away, there will be no picture of her smile. This is the essence of the quantum Zeno effect.

Why should this help settle the argument between "mind first" and "matter first"? It brings "I" back into the equation. The quantum Zeno effect shows that reality is like a bride who smiles naturally only as long as a camera isn't pointing at her. She doesn't like being looked at. But there's the rub. *We are always looking at reality.* There is no such thing as looking away. Which means that how the universe behaves when no one is looking has no meaning. (Of course, since human beings have only been around for a fraction of the life of the universe, it remains an open question as to what an observation really is and, by implication, who is doing the observing. For many physicists, there can be no observer who isn't human. We will return to this point later.)

The "matter first" camp refuses to accept this inescapable fact about constant observation. They are like a wedding photographer who says to the bride, "I don't care if you can't smile when a camera is on you. I'm going to keep the camera on you until I catch a smile." He can wait forever. Apparently so can the "matter first" camp, despite the quantum Zeno effect. It tells us that we will never see a particular molecule undergo a transition as long as we insist on looking at it. In fact, the more observations you make, the more frozen the unstable system will be.

So it must follow that the more we look at the world, and the closer we get to its finest structure, the more we are freezing it in place. Somehow observation gives specificity to reality. Reality slips through Sherlock Holmes's magnifying glass just when he thinks he's spotted a clue. But before the "mind first" camp starts to cheer, the quantum Zeno effect has bad news for them, too. There is no separate observer. The "matter first" people are stuck because they can't report what a physical system is doing when it behaves naturally. The "mind first" people are stuck because they can't produce an independent observer. The so-called observer effect only works if an observer can stand outside the system he wants to observe.

You can chop the observer up, as it were, by asking him to take a measurement of one small thing, such as detecting a photon as it passes through a slit. If you watch all the time, however, the observer has no way to step back from the thing he's watching. This is why the quantum Zeno effect is sometimes called the watchdog effect. Imagine a bulldog chained to the back door of a house. The dog has been trained to keep its eyes on the back door constantly, and to bark if anything suspicious happens. Unfortunately, the bulldog is so fixed on guarding the back door that burglars can sneak in the front door or a side window or anywhere else they please. You might as well not have a watchdog. In the same way, any observation made in physics locks the observer's attention on a single thing. As long as the two are locked, anything else could be happening all around and no one would know it. You might as well not have an observer.

This lock between observer and observed lies at the heart of the quantum Zeno effect. How can we break the lock? There's a good deal of controversy over that. Maybe the lock can't be broken. Maybe it can be broken through an equation but not in real life. Amidst all this speculation, something wonderful has occurred. Reality has spoken for itself, which was just the thing we needed. Reality's message is intimate: "I have you in my embrace. We are locked together, and the more you try to break away, the tighter my embrace becomes."

In other words, "matter first" and "mind first" must both surrender to "reality first." The observer has nowhere to stand outside reality. He's like a fish who wants to escape the sea only to find that if it jumps out of the water, it perishes. For human beings, participating in the universe is how we exist. To exist is to be aware. That's the long and the short of it for human beings. Astonishingly, the same is true for the universe. Without consciousness, it would vanish in a puff of smoke, like a dream, leaving nothing behind and no one to know that it ever existed. Even to say that the universe is conscious doesn't go far enough. As we will convincingly argue, the universe is consciousness itself. Until that conclusion is accepted, reality's message hasn't been completely heard.

HOW DID LIFE FIRST BEGIN?

Shakespeare has an unsettling habit of mixing nobility with tomfoolery, so that mad King Lear shaking his fist at a thunderstorm has no companion in the pouring rain but the poor fool who served him at court. A grinning death's head is always around the next corner in *Hamlet*. Hamlet utters high-flown sentiments like "What a piece of work is a man, how noble in reason, how infinite in faculty!" Meanwhile, the First Grave-digger (sometimes listed as First Clown) cracks jokes about how fast a corpse will decay when the ground is wet, including the corpses of great men. His jesting throws Hamlet into a morbid mood. In the end, what good are noble thoughts? he asks: "Imperious Caesar, dead and turned to clay, / Might stop a hole to keep the wind away."

In science, physics is Hamlet and biology is the First Grave-digger. Physics expresses itself in elegant equations while biology deals with the messiness of life and death. Physicists dissect space-time; biologists dissect flatworms and frogs.

For a long time physics wasn't concerned with the mystery of life. Erwin Schrödinger wrote a small book titled *What Is Life?*, but his colleagues generally viewed it as an eccentricity, a piece of mysticism rather than science—at least not the science of relativity and quantum mechanics, which was the business Schrödinger was trained to attend to. Actually, he was trying to connect genetics with physics, but at that time, 1944, the structure of DNA was still unknown. Even after the discovery of the double helix in the following decade, physics remained aloof from biology, a situation that has changed only gradually in the last few decades.

Equations and theories, scientific data and results, are faraway things; life is with us here and now. One of the most peculiar things about being alive is that we don't know how it happened and when it did happen. If you look at any living thing—a cold virus, *T. rex*, tree fern, or newborn baby—it was preceded by another living thing. Life comes out of life. Clearly this doesn't tell us where life first began, and yet the transition from dead matter to living matter somehow occurred. In biochemistry this pivotal moment is explained by setting inorganic chemicals on one side and organic chemicals on the other. An organic chemical is defined as a chemical that appears only in living things—organisms. Salt is inorganic, meaning it is not based on carbon, for example, while the flood of proteins and enzymes manufactured by DNA is organic.

But it's not clear that this time-honored division really helps if you want to know how life first began. The separation of organic and inorganic chemicals is valid as chemistry but not as a definition of life. Some amino acids, the building blocks of proteins, may be present on the surface of meteorites. In fact, one theory about the origin of life holds that the first spark came from such meteorites landing on Earth.

To be brutally frank, life is a major inconvenience for physics. Biology doesn't fit into abstract equations. If you consider what the experience of life feels like, even biology may be inadequate to explain it. Life contains purpose, meaning, direction, and goals—organic chemicals do not. It isn't tenable that chains of proteins somehow looked around and learned to do the things associated with living organisms. That's like saying that stones in a New England field looked around and decided to become a Yankee farmer's fence. And even if salt is "dead," life cannot exist without its participation—every cell in the body contains salt as a necessary chemical ingredient.

The fact that life comes from life implies that living things want to keep going. Unless extinction becomes total, evolution is apparently an unstoppable force, but why? Eons ago—to be precise, some 66 million years ago—we are told, a giant meteor struck Earth and wiped out all the dinosaurs, probably because the collision created so much dust in the atmosphere that sunlight was blocked and the planet became too cold for dinosaurs to survive, or else because plant life withered away and the entire food chain collapsed, making the survival of very large creatures impossible. From this mass extinction, the creatures that survived, tiny and insignificant as they were, didn't remain tiny and insignificant. The age of the mammals became possible. A new blossoming took hold, and the post-dinosaur world now looks far richer and more diverse than what came before.

The surge of life is both obvious and mystifying. The blue-green algae that form on the surface of ponds haven't evolved for hundreds of millions of years; neither have sharks, plankton, horseshoe crabs, dragonflies, or a host of other life-forms that lived alongside the dinosaurs. What causes some creatures to stay put while others gallop ahead on the evolutionary track, as pre-hominids did, creating *Homo sapiens* in record time, a matter of 2 or 3 million years instead of tens or hundreds of millions?

It's an axiom in science that the relevant questions are about "how," not "why." We want to know how electricity works, not why people want bigger flat-screen televisions. But the evolution of life keeps bringing up issues of why. Why did moles abandon the light to live underground? Why do pandas eat only bamboo leaves? Why do people want children? Some kind of purpose and meaning had to enter the picture. Or did a conscious universe contain the seeds of purpose and meaning since the beginning? As matters stand, such speculation is met with considerable resistance by the scientific community. The standard view holds that the universe has no purpose or meaning. So before offering a new model for how life began, we must dismantle conventional thinking first. In a conscious universe, everything is alive already. The observation that life comes from life turns out to be a cosmic truth.

GRASPING THE MYSTERY

The chemicals in the human body are the reason the body is alive. At the head of all organic chemicals is one, DNA (deoxyribonucleic acid), which contains the code of life. Yet if you stand back, this seems like an awkward, perhaps infeasible, way to unravel the mystery of where life began. Carbon, sulfur, salt, and water are supposedly dead, while at the same time being totally necessary to life, so why should organic chemicals be considered privileged?

What any living thing does, whether it's a microbe, butterfly, elephant, or palm tree, isn't the same as what it's made of. No shuffling around of chemicals will cause a piano to write a piece of music. Like the human body, the wood that encases a piano is composed entirely of organic chemicals, primarily cellulose. Nothing about cellulose explains the music of the Beatles, or any other. Likewise, jiggling around the chemistry of the human body doesn't explain any living activity a person performs. Genetics would seem to be on wobbly ground.

You might make a special plea for the chemicals in the human body as opposed to the lifeless chemicals in seawater and a piece of wood, but there will always be a hidden fallacy, a weak link that snaps. One way to illustrate this is through an aspect of every living cell known as nanomachines, microscopic entities that function like production plants to manufacture the chemicals a cell needs in order to survive and multiply.

Our cells don't need to reinvent the wheel. DNA isn't made from scratch every time a new cell is created. Instead, DNA splits itself in half in order to form a mirror image of itself, and that becomes the genetic material for a new cell. (How this act of self-replication comes about has no explanation, but we'll leave this mystery aside.) The cell doesn't want to make other chemicals from scratch, either. Evolution has led to a host of fixed machines that persist intact during the life of a cell. They are like coal and steel plants that never close down or get dismantled no matter how much change occurs in the city around them. A particular zone in the cell, known as the mitochondrion, which provides the energy for the cell, is so stable a nanomachine that it gets passed on unchanged generation after generation. You inherited your mitochondrial DNA from your mother, and she from her mother, as far back as human evolution can be traced. In one form or other, the mitochondrion has been stable in every living cell as its energy factory. The traffic of air and food inside a cell is constantly swirling and changing, but nanomachines are immune to this traffic. In fact, they guide it in many ways.

THE MACHINERY OF LIFE?

If we want to get at the very beginning of life, nanomachines sit at the very heart of the mystery. But first, like Alice, we have to go through the looking-glass into a world where the tiniest things, atoms and molecules, loom large. They are in control of reality at the microscopic level. Whatever happens in nature, whether in the center of a supernova, the gas clouds of deep space, or a living cell, is happening through the interaction of atoms and molecules. Nothing else is germane to how life began in material terms. If atoms and molecules cannot accomplish the job on their own, it can't be done. This is what current biology holds. For the moment we'll exclude quanta, although we'll return to them later.

Atoms interact with each other almost instantaneously. You may have heard of chemicals known as free radicals that exist in the human body, being involved in many processes both destructive and constructive. Free radicals therefore are double-edged swords; they are associated with aging and inflammation, for example, yet at the same time they are necessary for healing wounds. The basic thing that free radicals do is quite simple, however—they steal electrons from other atoms and molecules. Their own count of electrons is unstable—because of exposure to radiation, smoking, and other environmental factors or from the body's own natural processes. The immune system creates free radicals to steal electrons from invading bacteria and viruses as a way of neutralizing them. The most common atom involved in electron stealing is oxygen. When its electron count becomes unstable, oxygen latches on to the nearest electron it can steal. Therefore, free radicals are highly reactive and usually very short-lived.

For living organisms and their cells, this is a life-or-death matter. It boils down to the paradox that life requires stability and instability at the same time. Life also requires that vastly different timescales, from nanoseconds to millions of years, somehow be tied together—a cell operates in thousandths of a second but took tens of millions of years to evolve.

The meshing of opposites that makes life possible isn't theoretical. Inside a cell, some atoms and molecules must be freed up to do various jobs by bonding with other atoms and molecules, yet, having done their job, stable substances must persist without ever changing. But which atom goes where? They don't come with address labels. To compound the problem, some of the most important organic chemicals, chiefly chlorophyll in plants and hemoglobin in red-blooded animals, carry the tricky balance of stability versus instability to amazing extremes.

Hemoglobin sits inside a red blood cell, constituting 96 percent of the cell's dry weight; its function is to pick up oxygen and transport it through the bloodstream to every cell in the body. Blood gains its red color from the iron in hemoglobin, which turns red after it picks up an oxygen atom, exactly as iron turns reddish when it rusts (and for the same reason). When the oxygen atoms reach their destination and are released, the red color fades, which is why blood in your veins is bluish. Venous blood is on the return journey to the lungs, where it will start the process of oxygen transport all over again. The ability of hemoglobin to carry oxygen is seventy times greater than if the oxygen were simply dissolved in the blood. (All vertebrates contain hemoglobin except fish, who pick up oxygen from water through their gills instead of breathing air and therefore employ a different process.)

As a molecule, hemoglobin is a miracle of construction. Since we've gone through the looking-glass, let's imagine walking into the hemoglobin molecule as if entering a vaulted building like a greenhouse with spidery chains of smaller molecules forming the girders and beams. At first it would be hard to even see the iron atoms that are the whole reason for hemoglobin's existence. Ribbons of proteins form helixes, and other chemicals link the helixes, functioning like welded bolts. With an eye for pattern, we discern that the protein chains hold a specific shape. There are subunits within the units or proteins, each bonded to the only thing that isn't a protein, the iron atoms formed into *hemes*—these are rings of proteins encircling the iron. In structural terms there are also specific folds and pockets that need to be in place.

Think of rich people living in huge mansions that rationally speaking are a waste of space for one or two people to rattle around in. The hemoglobin molecule is built from 10,000 atoms, creating a vast space that exists so that exactly four iron atoms can pick up four oxygen atoms for transport. These 10,000 atoms aren't some kind of luxurious waste, however; they are recombinations of simpler proteins also necessary for the life of cells. Besides containing hydrogen, nitrogen, carbon, and sulfur, the structure of hemoglobin contains oxygen. So the actual task that faced inorganic matter billions of years ago on planet Earth was as follows:

> Oxygen had to be set free into the atmosphere without getting gobbled up by greedy atoms and molecules around it.
>
> At the same time, some of the oxygen had to be gobbled up to form complex organic chemicals.
>
> Those organic chemicals had to be structured into proteins, of which hemoglobin is one of the most complex.
>
> Hemoglobin had to be arranged internally so that it encased four iron atoms, which are absent from hundreds of other proteins, including those that resemble hemoglobin in their working parts.
>
> The iron atoms couldn't be inertly encased, like locking diamonds up in a safety deposit box. The iron had to be charged (as a positive ion) so that it could pick up oxygen atoms. But it wasn't permitted to steal any of the oxygen already being used to build proteins.
>
> Finally, the machinery necessary for constructing all of the above organic chemicals had to *remember* how to do it the next time and the next and the next, while other nanomachines sitting nearby in the cell had to remember hundreds of different chemical processes without interfering with the machine that makes hemoglobin. Meanwhile, in the nucleus of the cell, DNA has to remember—and put into motion with precise timing—the whole enterprise.

No matter how you cut it, this is a lot to ask of atoms, whose natural behavior is to bond instantaneously to the atom next door and stay that way. And this natural behavior hasn't gone out of fashion; the countless sextillions of atoms in stars, nebulae, and galaxies are acting as they always have. So are the atoms contained in the solar system, the sun, and our planet—aside from the atoms in living creatures. Those atoms manage the trick of behaving naturally while at the same time pursuing a creative sideline, namely, life.

As animal life was humming along creating hemoglobin, natural processes on the vegetable side of the operation created chlorophyll, which sustains plant life along a different route, photosynthesis. We won't conduct a tour of the chlorophyll molecule except to say that it consists of 137 atoms, whose sole purpose is to encase one atom of magnesium rather than the iron in hemoglobin. This ionized magnesium atom, when it comes into contact with sunlight, allows carbon and water to form a very simple carbohydrate. How photons of light from the sun can create this new product opens up new mysteries, but once the simplest carbohydrate molecule was generated by plant leaves, an evolutionary breakthrough was made. The machinery that manufactures chlorophyll took a separate track from the machinery that manufactures hemoglobin, which is why cows eat grass instead of *being* grass.

(Note: In photosynthesis, chlorophyll only needs the carbon atom in carbon dioxide, releasing the oxygen atom into the air. You may say, aha, that's where the free oxygen comes from that isn't stolen by other atoms. But unfortunately, chlorophyll needs a cell to live inside, and that cell required free oxygen for its construction before chlorophyll could start to operate.)

Now we have a context for asking the right question. The mystery of how life first began comes down to the transition from "lifeless" chemical reactions to "living" ones. Is life simply a sideline of universal chemical behavior throughout creation? Any answer will also have to include why only some atoms and molecules engage in this sideline while the rest continue on their merry way.

THE JOURNEY FROM SMALL TO NOTHING AT ALL

It turns out that getting around "life comes from life" is no easy feat. Absolute beginnings don't seem to exist. But the urge to go smaller and smaller is irresistible to scientists. The oldest living things were microscopic in size, much smaller than cells, which didn't evolve until hundreds of millions of years later. The most recent finds indicate that 3.5 billion years ago, only a billion years after Earth was formed, complex microbial life had already taken hold. There may be fossils of bacteria detectable in very old rocks, as some microbiologists believe. But every time one is discovered and dated, it gets challenged. It's extremely difficult to know if you're looking at a fossil or the traces of a crystal.

Perhaps the secret lies at a level even smaller than bacteria and viruses, so we could knock on the door of molecular biology, the field that has revealed everything we covered about hemoglobin and chlorophyll. The scientist who answers the door would only shake his head if we ask where life came from. "The organic chemicals I study already exist in living things," he says. "No one knows where they originated. Chemicals don't leave fossils."

We could remind him that evidence of amino acids has been found on meteorites. Other people speculate that life may have existed on Mars before it evolved here on Earth. If a big enough asteroid hit Mars, it could have blown chunks of rocks into space, and if one of those arrived on Earth and the life sticking to it survived the journey through outer space, maybe that's how organic chemicals began here.

Our molecular biologist offers a dismissive remark as he shuts the door. "These kinds of speculations are closer to science fiction than science. They have no evidence to back them up. Sorry."

And so it goes, like a bad dream in which an endless corridor leads to one door after another, endlessly. No matter how small you reduce the problem, there is always a smaller level, until the whole thing—matter, energy, time, and space—vanishes into the quantum vacuum and leaves us with a very frustrating situation, because there *has* to be an answer—after all, life is here, all around us. The journey from living things that takes us to nothing must be reversible. "Life comes from life" doesn't let us off the hook for explaining how, to begin with, life entered the picture.

In a curious and very clever way, one of the originators of the multiverse, physicist Andrei Linde, uses nothing to show why human life must have come about. When asked about the most important recent discovery in physics, Linde picked "vacuum energy." This is the finding that empty space contains a very tiny amount of energy. We've touched upon this fact, but Linde works it into the reason for life on Earth.

At first glance the amount of vacuum energy looks quite trivial. "Each cubic centimeter of empty interstellar space contains about 10^{-29} grams of invisible matter, or, equivalently, vacuum energy," Linde points out. In other words, invisible matter and vacuum energy are fairly comparable. "This is almost nothing, 29 orders of magnitude smaller than the mass of matter in a cubic centimeter of water, 5 orders of magnitude smaller than the proton. . . . If the whole Earth would be made of such matter, it would weigh less than a gram."

The importance of vacuum energy, tiny as it is, was vast. The balance between the energy in empty space and the invisible matter in empty space gave us the universe we inhabit. Too much of one or the other, and the universe would either have collapsed upon itself soon after the big bang or would have flown apart into random atoms that never gathered into stars and galaxies. Here is where Linde finds the key to life on Earth.

Vacuum energy isn't constant, he believes. As the universe expands, the density of matter will thin out as galaxies fly farther and farther apart. As this happens, the density of vacuum energy will also change. Somehow, human beings happen to live at the perfect point of balance—and we must live there. We sprang up—life sprang up—at a place that has to exist. Why? Because as vacuum energy is tipping the scales one way or the other, all possible values come about. One might imagine a family's home movies of the kids growing up. Most of the movies got lost accidentally, but there's footage of one baby being born and then the same child at age twelve. Even with missing footage, it must be true that every stage of growth between one day and twelve years existed.

Linde's origins story for life on Earth is the best that anyone has to offer, he says, and the story takes an optimistic turn. "According to this scenario, all [vacuums] of our type are not stable, but metastable. This means that, in a distant future, our vacuum is going to decay, destroying life as we know it in our part of the universe, while re-creating it over and over again in other parts of the world."

Sadly, there's a fly in the ointment. "Metastable" means that areas of instability get canceled out if you stand far enough back. The carbon inside a dying person's body is just as stable as the carbon in the body of a newborn baby. Standing back, nothing counts that happened between birth and death. That's fine for chemistry class but useless in real life. The vacuum state is stable as galaxies are born and die, or as the human race emerges and then meets extinction. This says nothing about where life came from, only that the stage was set for it. Linde does an elegant job of setting the stage—perhaps the most elegant job anyone has ever done so far—but he doesn't take us from nothing to the origin of life.

ARE QUANTA ALIVE?

The multiverse hasn't really solved the mystery of life, and there's a better clue, which relates to ordinary energy, like heat and light, rather than the exotic species of vacuum energy. The behavior of ordinary energy is to even out, so when energy starts to clump up, it immediately tries to escape the clump and reach a flat state. That's why a house where the furnace goes out in winter gets colder and colder until it's the same temperature inside and outside. The heat evened out.

This dissipation of energy is known as entropy, and all life forms resist it. Life consists of energy clumps that do not even out until the moment of death. When you wait for the bus in winter, unlike a house where the furnace burned out, your body remains warm. This isn't because you are well insulated by wearing a thick coat against the cold. Instead, your body extracts heat energy from food and stores it at a constant temperature, around 98.6 degrees Fahrenheit. Every schoolchild is taught this fact, but if we knew how organisms first hit upon the trick of defying entropy, that might be why life exists in the first place.

Almost all the free energy available for life on our planet comes from photosynthesis. Besides needing their own supply of energy to grow, plants are at the bottom of the food chain for all animal life on land. When sunlight hits cells that contain chlorophyll, the energy in the sunlight is "harvested," almost instantaneously being passed along for chemical processing into proteins and other organic products. This energy transfer occurs almost instantaneously and with 100 percent efficiency. No energy is wasted as heat. By comparison, if you go out for a morning run, your body's efficiency at burning fuel leads to a lot of excess heat as you sweat and your skin gets warm; there is also much chemical waste that must be carried away from your muscles in the bloodstream.

Chemistry couldn't explain the near-perfect precision of photosynthesis. In 2007 a breakthrough was made at the Lawrence Berkeley National Laboratory by Gregory Engel, Graham Fleming, and colleagues, who came up with a quantum-mechanical explanation. We've already covered that photons can behave like either waves or particles. The instant a photon makes contact with the electrons orbiting in an atom, the wave "collapses" into a particle. This should lead to a lot of inefficiency in photosynthesis. Like shooting darts at a board, there will be a lot of misses before the bull's-eye is hit. But the Berkeley team discovered something quite unique: in photosynthesis sunlight retains its wave-like state long enough to sample the whole range of possible targets while simultaneously "choosing" which one is the most efficient to connect with. By looking down all the possible energy pathways on offer, the light won't waste energy picking any but the most efficient ones.

The details of the Berkeley findings are complex, centering on long-term quantum coherence, which means the ability of the wave to remain a wave without collapsing into a particle. The mechanism involves matching the resonance of both the light and the molecules receiving its energy. Think of two tuning forks vibrating exactly alike; this is known as harmonic resonance. At the quantum level, there is a similar harmony between the oscillations of certain frequencies of sunlight and the oscillations that the receiving cells are tuned to.

Quantum effects are known to exist in other key places where micro meets macro. Hearing is stimulated in the inner ear by oscillations that are quantum in scale, being smaller than a nanometer (i.e., a billionth of a meter). The nervous systems of some fish are sensitive to very small electric fields, and our own nervous system generates very tiny electromagnetic effects. The exchange of potassium and sodium ions across the membrane of each brain cell gives rise to the electrical signals transmitted by the cell. An entire new theory posits that living things are embedded in a "biofield" that originates at the electromagnetic level or perhaps at an even subtler quantum level, yet to be explored. As you can see, quantum biology has a real future. The breakthrough with photosynthesis was a turning point.

Yet as intriguing as all of these discoveries are, declaring that quanta are alive won't tell us how they acquired life. The snake just winds up biting its tail again. If human beings are alive because quanta behave in a totally lifelike way (i.e., making choices, balancing stability and spontaneity, efficiently harvesting energy, and so on), all we've proved is that life comes from life. This is something we already knew.

Quantum effects in biology are important, nevertheless, because they introduce behavior that isn't predetermined the way oxygen atoms are when reacting with other atoms. A word like *choice* implies that determinism has been loosened up a bit. But is this enough? As green leaves flutter in the trees, sunlight is used to build a carbohydrate thanks to a quantum decision, yet this isn't enough to tell us about the decisions being made up the line, where a single liver cell performs dozens of processes in coordinated fashion with trillions of other cells. In building a house, knowing how to mortar each brick is important, but it's not the same as designing and constructing the whole house.

GETTING FROM "HOW" TO "WHY"

With science stymied to explain how life originated, maybe we've been asking the wrong question. If someone throws a brick through your window at midnight, you can't see who did it in the dark. But that's secondary to asking why they did it. Clearly our lives have purpose, while nature, we are told, has no purpose—it just is. Being without purpose brings no sleepless night to quarks, atoms, stars, and galaxies. Why go off on a tangent and create living organisms that are driven by food, mating, and other reasons to be alive?

We believe that the absence of purpose is inconceivable. As long as you are human, A leads to B for a reason. There is no other way to use the brain. Without purpose, no events exist, at least not as perceived through a human nervous system. Let's say that you've been marooned on a desert island for sixty years. One day out of the sky a package parachutes to earth, and when you open it, there are two objects inside, a smartphone and a desktop computer. Both run on batteries. It wouldn't take you long to figure out that the smartphone, even though it looks nothing like the telephones you knew from the 1960s, works as a telephone. Because you know why it exists, you have a fairly easy path to using it. You wouldn't need to know how the smartphone worked once you made the connection between punching in numbers and hearing a voice answer at the other end.

But the computer is another story, because in the world you left behind around 1965, computers were in their infancy, and nothing about a desktop computer looks like the massive IBM mainframes you saw on television. Fiddling around, you would need hundreds of hours to figure out by hit and miss what you're dealing with. This strange machine isn't the same as a typewriter or a television, even though it has both a keyboard and a screen. Let's say you are mechanically inclined, and you are able to open up the inner workings of the computer. Inside you see a wealth of parts that make no sense to you. Is it conceivable that on your own you could grasp how a microchip works? Even if you did, would that information tell you how to run the computer's software?

The answer is most likely to be no on all counts. Unless you know why a computer exists, the same way you know why a telephone exists, taking apart the machinery won't get you from *how* to *why*. Many airline passengers don't know how an airplane manages to fly, but they get on board because they need to travel somewhere; the *why* of the airplane is enough. A plane exists to take you places faster than a car or train. So why does life exist? It certainly doesn't need to. All of the chemical components and quantum processes that interact to create life were sufficient on their own already.

Like Frankenstein's monster being jolted with electricity from a lightning storm, it would be very helpful if some basic physical trigger—the spark of life—automatically made life happen. But no such trigger exists. Looking out over the vast panorama of living organisms, we are stuck with the undeniable fact that life always comes from life, not from dead matter. Even in laboratories where new forms of bacteria are being engineered, so-called artificial life is still a recombination of DNA being sliced and diced. (If the manufacturer wants to design a specific micro-organism that feeds on petroleum, which would be very useful in cleaning up oil spills at sea, devising this new life-form has a chance of success only by working from preexisting organisms that feed off oil in some form. Without a goal in mind, tinkering around with DNA basically goes nowhere.)

Nature, however, wasn't so lucky. It had to build living organisms blindly, without knowing in advance what needed to be built. Nature wouldn't even know if it made a mistake along the way, because unless you know where you're going, no choice is right or wrong.

Billions of years ago, oxygen atoms had no inkling that life was around the bend. No one told them that sunlight was going to be harvested, or that they'd be necessary in organic chemistry. Life brought about huge adaptations on our planet, and yet oxygen atoms don't adapt. Most scientists would shrug their shoulders and insist that blind nature created life through automatic, deterministic processes. The bonding of atoms leads to simple molecules; the bonding of simple molecules leads to more complex molecules; when these molecules are complex enough, life appears. As far as mainstream science is concerned, this wholly unsatisfactory story is basically all there is.

To arrive at a better story, we must explain *why* life was needed in a system—planet Earth—that was perfectly sufficient without it. Knowing the how isn't useless; we're not claiming that. But imagine that you want to buy a house. You go to the bank, and the loan officer presents you with a stack of papers to fill out. He explains that each piece of paper is necessary. You can't skip any, and if your application is found wanting at any step of the way, the deal collapses. Millions of people have gritted their teeth and filled out every piece of paper for one and only one reason: they want a house. Having a goal in mind, they are willing to endure the necessary steps to get there.

Nature had to go through thousands of linked steps in order to produce living organisms. Do we really buy the story that this happened without a goal? It's as if a customer came into the bank, filled out dozens of forms at random, and one day was told, "You own a house. We know you didn't come in for one, and you had no idea what those pieces of paper were for."

Now we know what is lacking if we want to understand where life came from. Without a why, the whole project is too incredible ever to happen. Knowing that life is the goal, rather than having to rely on random change, would make everything a thousand times easier to explain. But suddenly a new mystery has opened up. If life was part of the cosmos from the start, what about mind? At the instant of the big bang, was the human mind inevitable? The reason we are forced to ask it is simple. Unless the universe is mindful, it's impossible to create mind out of a mindless creation. As Sherlock Holmes liked to remind Watson, once you've eliminated every other possible solution, the one that remains must be true. In this case, a universe that is thinking all the time sounds incredible, but every other answer, as we will see, turns out to be wrong.

DOES THE BRAIN CREATE THE MIND?

Before the universe can have a mind, we need to understand our own minds. That's logical enough. We cannot see reality through the minds of dolphins and elephants, even though both species have outsize brains that could be functioning at a very high level. Almost certainly there is a dolphin reality and an elephant reality, tailor-made to fit their nervous system. Dolphins have been shown to learn words, giving them a close affinity to humans, and also, like humans, they are capable of savage acts. Still, they are not humans, inhabiting a reality beyond ours.

This logic leads to a surprising conclusion. A universe is defined by the creatures who inhabit it. What humans call "the" universe is like taking two bananas, a sack of flour, and a frozen pizza home and claiming that you bought out the supermarket. Any reality, as perceived through a different nervous system, implies a different universe, so that dolphins and elephants inhabit their own, which to them is "the" universe. And why stop with them? Why not a snail universe or a giant-panda universe? Humans didn't take out an exclusive deed on reality—we just assume we did, perhaps out of our self-inflicted sense of superiority.

The reason we made that assumption is pride of the brain. With its quadrillion possible combinations, the human brain is the most complex object in the universe, so far as we know. Thanks to its activity, we are self-aware. A horse eats grass and is content. We eat spinach and can say, "I don't like this" or "I love this," along with any opinion in between. This implies enormous control over our thoughts. Pride of the brain lies behind all of science, too, since our brain has a mysterious capacity for logic and reason (the newest abilities early man acquired, brainwise, when the cerebral cortex evolved, not in terms of millions of years like the lower brain but perhaps tens of thousands). Pride of the brain is seriously humbled, however, when we look more closely.

First of all, science, at least classical physics, is in love with predictability, but our minds aren't. One of the easiest bets to win is to offer a million dollars to anyone who can accurately predict their next thought. It would be foolhardy to accept such a wager. As we all experience every day, our thoughts are unpredictable and spontaneous. They come and go at will, and strangely enough, we have no model for how this works. The brain is supposedly a machine for thinking. But what kind of machine churns out so many different responses to the same input? It's like the world's most dysfunctional candy machine. You put in a nickel, but instead of getting a gumball every time, the machine spits out a poem or a delusion, a new idea, or a trite cliché, occasionally a great insight or a bizarre conspiracy theory.

One theory of mind and brain actually does recognize the unpredictability of thought and links it to the quantum realm. Roger Penrose, working in collaboration with anesthesiologist Stuart Hameroff, departed from the conventional notion that consciousness is produced by activity occurring in the synapses, the gaps between brain cells. Their theory, known as Orchestrated Objective Reduction (Orch-OR) looked instead to quantum processes that happen inside the neuron. In other words, the *reduction* in Orch-OR's title is drastic, examining much finer fabrics of nature than chemical reactions. In a microscopic structure of cells known as microtubules, Penrose and Hameroff propose that unpredictable activity at the quantum level is the origin of events happening in consciousness. Mind needs the quantum to exist.

The other two words in the title are just as important. *Orchestrated* means that orderly brain activity is being controlled from the very origins of the brain at a microscopic level. This is attractive because a basic quality of consciousness is orderly, organized thinking. *Objective* is important because scientists want to preserve the assumption that anything in creation, including consciousness, must be explainable by physical (i.e., objective) processes. In our view, this assumption falls apart when it comes to the inner world of human experience. We don't accept that mind needs the quantum. Penrose and Hameroff took a bold step delving into quantum biology, and it is likely that future theories, or a future revision of Orch-OR, will continue to examine the brain at this level.

From our perspective, one particular advantage of Orch-OR is its assertion that the human mind cannot be computed through mathematical formulas. In other words, no matter how predetermined the firing of a neuron is, the thoughts that neurons process are not predetermined. Hameroff and Penrose arrive at this conclusion through some sophisticated quantum reasoning, along with hints from philosophy and advanced logic. But the upshot is fairly simple: no mechanical model will ever explain how humans think. A great deal of confusion and inevitable dead ends would be avoided if other scientists took this point to heart.

Like it or not, our minds are on dual control. Sometimes we are the ones in control. Sometimes a totally unknown force is in control. This isn't hard to see. If you are asked to add 2 + 2, you can call up the necessary mental process to arrive at the right answer because you are in control. There are millions of similar tasks, such as knowing your own name, how to do your job, what it takes to drive a car home from work—and these give us the illusion that we control our own minds all the time. But someone suffering from anxiety or depression is the victim of uncontrolled mental activity, and lack of control can go much further, as in mental illness, for example. A common symptom of various psychoses, particularly paranoid schizophrenia, is the belief that an outside agent is controlling the patient's mind, usually through an alien voice heard in the head. A normal person doesn't usually feel out of control mentally, but if it were really true that we have control over our thoughts, we'd call up any thought we wanted to have, the way you can call up a Google search, and this is far from the case.

Love at first sight is a pleasant way to be out of control, and so is the experience of artistic inspiration. We can only imagine the joy of Rembrandt or Mozart in the throes of creating a masterpiece. So dual control has its good and bad side. Life would be robotic if we didn't have flashes of emotion that come of their own accord, along with bright ideas of every kind. What if this everyday fact of life turns out to be the key to the cosmos? Human beings might be a bright idea the universe had, and once the idea occurred to it, cosmic mind decided to run with it. Why? What's so enticing about human beings, troublesome and pained as we are? Only one thing. *We allowed the universe to be aware of itself in the dimension of time and space.*

In other words, at this very moment, the cosmos is thinking through you. Whatever you happen to be doing—riding a bike, eating a Reuben sandwich, making a baby—is a cosmic activity. Take away any stage in the evolution of the universe, and this very moment vanishes into thin air. As astounding as such a claim may be, this book has been building up to it all along. Quantum physics makes it undeniable that we live in a participatory universe. Therefore, it's only a small step to say that the participation is total. Our minds are fused with the cosmic mind. The only reason it took so long to arrive at this conclusion is that old bugaboo—stubborn materialism. As long as you look upon the brain as a thinking machine, there cannot be a cosmic mind, because in physicalist terms, no brain = no mind. The obstacle couldn't be more intractable.

To remove the obstacle and allow the human mind to fuse with cosmic mind, we must address the mystery of how the brain is related to the mind. There's no way around it. The first person who called the human brain "the three and a half pound universe" created an indelible image. If the brain is a unique physical object that functions like a supercomputer, then the physicalists have won. But there is no reason to elevate the atoms and molecules inside our brains to special status. If every particle in the cosmos is governed, created, and controlled by the mind, the brain also functions as the mind dictates. That's the key to solving this, our last mystery.

GRASPING THE MYSTERY

It's amazingly difficult to figure out what the brain actually does. If nature has a sense of humor, this is the ultimate prank, keeping the brain under wraps even though the mind is using it at every moment. You can't figure out how a neuron works simply by thinking about it; indeed, you can't even figure out that neurons exist. We don't see or feel our brain cells. With the dawn of X-rays, fMRI scans, and sophisticated surgical techniques, neuroscience can make the brain's machinery visible. There it sits, twinkling away with microvolts of electricity, zapping a few molecules of neurotransmitters across the synapses, and yet for all intents and purposes, brain cells act like all the other cells in the body. Even skin cells secrete various neurotransmitters. So why do you have to open your eyes to see the sunrise instead of just holding up your elbow to see it?

No one has managed to close the gap between what a brain cell does (bouncing atoms and molecules around) and the rich four-dimensional world that the brain manages to produce. To get around this fundamental difficulty, reality has to be rethought from the ground up. Equating the brain with a computer is a common assumption that can be thrown out almost immediately. Let's say you saw a lovely pink rose named Queen Elizabeth and decide to plant one in your garden. When you arrive at the nursery, the rose's name skips your mind, but after a moment you recall it. If instead you asked your smartphone to find the right name, it would go through every single pink rose in its memory chips, and as it undertook this laborious process, it would never know that Queen Elizabeth was the correct name until you told it so.

Computers are in no way smart. Worldwide publicity surrounded the 1997 defeat of the reigning world chess champion Garry Kasparov by a computer program from IBM known as Deep Blue. The two, man and machine, had been exchanging wins and losses for two years, and Deep Blue's ultimate victory was hailed as a step forward for artificial intelligence. But that's exactly the point: what the computer did was artificial. In a sophisticated software program that IBM continually refined and updated, the basic operation was to comb through every possible chess move in order to arrive at the one that was statistically likely to be the best. So in a sense, Kasparov versus Deep Blue was a contest between humans on both sides, but with vastly different approaches.

A human chess player doesn't remotely follow this procedure. Instead, the skill of playing chess has been mastered, and with this mastery comes a sense of strategy, imagination, and the ability to assess one's opponent—and many wins are due as much to psychological mastery as to skill. A champion "sees" the right move without running through all possible moves. Deep Blue in fact couldn't play chess in the first place; it could only run numbers and play the odds. The main reason this roundabout strategy worked in the end was that the programmers resorted to shortcuts that imitated how the human mind works, but the computer had no way of coming up with such shortcuts on its own. In effect, calling Deep Blue intelligent is the same as calling an adding machine intelligent—and equally off the mark.

Likewise, human beings experience a world of inner experiences like love, joy, inspiration, discovery, surprise, boredom, anguish, and frustration that cannot be turned into numbers. Therefore, the whole inner world is alien to computers. Hardcore AI experts tend to dismiss the inner world as a kind of glitch or even an illusion. If so, that would make the entire history of art and music an illusion, along with every act of imagination and all emotions and in the final analysis, science itself, since science is also a creative process. Clearly the mind can't be digitized; therefore, turning the brain into a supercomputer is fallacious, since everything it does is digitized.

Five Reasons Computers Are Mindless

Minds think. Computers massage digits.

Minds understand concepts. Computers understand nothing.

Minds worry, doubt, self-reflect, and await insight. Minds have feelings. Computers spit out answers based on crunching numbers.

Minds ask why. Computers ask nothing unless ordered to by someone with a mind.

Minds navigate through the world by having experiences. Computers have no experiences. They run software, nothing more, nothing less.

In fact, the computer model of the brain has risen to prominence only because previous models proved to be so inadequate. We can briefly tour the junkyard of rusted-out models, noting along the way why each contains a fatal flaw as it attempts to explain the mind as a brain operation.

Denial: This is ground zero, the claim that only the brain exists, the mind being a by-product that has no reality on its own. Deniers have one great advantage: they can go about business as usual without bothering with the mind. That's an appealing prospect to many. After all, they say, the practical business of science doesn't need to talk about the mind; it needs to do experiments and collect data. There's also "soft" denial, which says that the mind exists, but it's a given, like oxygen in the air. Both are necessary, yet you can do science for a lifetime without needing to refer to them.

The fatal flaw in denial: Deniers can't explain many things, notably the mind-like behavior of quantum particles and the observer effect (see page 19). The fact that consciousness changes the quantum world is just as practical as any other scientific fact. Therefore, leaving mind out of the argument isn't viable. There's also the unavoidable way that mind and matter are constantly interacting in the brain. Thoughts give rise to chemicals, and vice versa. No one could seriously call that unreal.

Passive perception: Another camp admits that the mind is real but limited. The brain knows the world through the five senses, acting as a data collector. This viewpoint is appealing because science itself is all about data. Like a point-and-shoot camera, the brain is passive but very accurate; it brings an object into focus, and trusting this picture, along with the other four senses, is good enough. If you need better data—something science can't do without—there are always better telescopes and microscopes to extend our vision to regions the eye alone cannot see.

The fatal flaw in passive perception: All the microscopes, telescopes, X-ray machines, and every other instrument created to act as passive perceivers do not perceive anything without the human mind to interpret them. The minds that constructed these devices didn't do it passively. The creativity of consciousness was involved, which goes far beyond mere data collection.

Complexity equals consciousness: This camp has an expanded view of the mind as a very complex phenomenon. In fact, complexity can help us understand how the primitive nervous system of worms, fish, and reptiles evolved into the infinite richness of the human brain. The appeal of complexity theory is that it skirts the thorny issue of how dead matter somehow "learned" to think and light up on a brain scan. Matter is matter, period. But over the course of billions of years, simple atoms and molecules evolved into incredibly complex structures. The most complex of these structures are associated with life on Earth. If life is the byproduct of complexity, then by the same logic the properties of living things can be traced to their complexity.

For example, one-celled organisms floating in pond water will seek the light, and from this primitive response, all visual systems evolved, including the eagle's eye, which can spot the movement of a mouse from hundreds of feet in the air. Likewise, everything the human brain can do has an ancestral origin in creatures that do it less well, the way chimpanzees use rudimentary tools and honeybees dance in a pattern that maps out where the best source of pollen is. In a world of ever-evolving complexity, the human brain sits at the apex as the crown jewel. Complexity gave the brain its abilities, including thought and rationality.

The fatal flaw in complexity equals consciousness: No one has ever shown how complexity explains the attributes of life. As we mentioned before, adding more cards to a deck doesn't mean that the deck will suddenly learn how to play poker. Taking primitive bacteria and throwing more molecules at them doesn't explain how the first cells came into being and certainly not how these cells learned complex behaviors.

The zombie hypothesis: This camp is marginalized but has gained media attention from its catchy name and the publicized efforts of one staunch advocate, philosopher Daniel Dennett. The basic premise is deterministic. Every brain cell operates by fixed principles of biochemistry and electromagnetism. Neurons exist without choice or free will. They are trapped and conditioned by the laws of nature.

Therefore, since every person is the product of brain cells, each of us is essentially a puppet dependent on physical processes we have no control over. Like zombies, we go through the motions of living entities, but our belief that we have choice, free will, a separate self, and even consciousness amounts to a reassuring story that we zombies tell around the campfire to keep warm. Akin to the complexity theory of mind, the zombie theory holds that consciousness is a by-product of the brain's quadrillion neural connections. Build a supercomputer with just as many connections, and it will be as seemingly conscious as a human being.

The fatal flaw in the zombie hypothesis: Two fatal errors come to mind (leaving aside the preposterousness of the claim that human beings aren't conscious, which smacks of mischief rather than serious thinking). The first flaw is creativity. Human beings are capable of practically infinite acts of invention, art, insight, philosophy, and discovery that can't be reduced to fixed cell functions. Second, the zombie argument is self-contradictory, because the people who espouse it, being zombies, have no way to show that their notions are trustworthy. It's like having a stranger come up to you saying, "I'm going to tell you all about reality. But first you need to know that I'm not real."

WHY YOUR BRAIN DOESN'T LIKE THE BEATLES

It's easier to kill a vampire with a stake through the heart than to dispel the assumption that the brain, a physical object, has the power to create the mind. But at least we've seen the fatal flaws in current theories of brain and mind. Dismantling a bad idea, however, isn't the same as finding a better one. We can unfold a better idea through Paul McCartney's beautiful singing of a Beatles classic, "Let It Be." Does your brain appreciate the song or does your mind? On the brain side of the argument, neuroscientists can pinpoint some specific brain processes when "Let It Be" enters the ear canal as sound vibrations.

Researchers at McGill University in Toronto hooked up subjects to electrodes that measured their brain activity as they listened to music. As one would predict, music creates its own pattern of response compared with nonmusical sounds. Raw input that reaches the auditory center in the cortex gets scattered into specific locations where rhythm, tempo, melody, tone, and other qualities are separately processed in a matter of milliseconds. The prefrontal cortex even compares the music you are hearing now to music you expect to hear from past experience. By comparing the two, your brain can be challenged by something it never expected to hear, and further, this can be a delightful surprise or a distasteful one.

The research also shows that the brain gets "hardwired" in childhood according to what system of music it's exposed to. A Chinese baby's brain develops specific connections that respond to Chinese harmony, thus leading to its enjoyment. A baby born in the West, exposed to Western harmonies, is hardwired to enjoy that system rather than the Chinese. Finally, the researchers could take a musical performance and gradually change it via computer software to see if the brain notices any difference.

Can you tell the real Paul McCartney from the best synthesized version? It depends. As the music becomes more mechanical and less personal, the brain often doesn't notice any difference until the change is glaringly obvious. This might explain "tin ears" and, at the other extreme, a professional musician's subtle ability to detect the finer points of musical style. Different wiring leads to different levels of appreciation.

The research on music and the brain has become quite sophisticated. Yet we'd argue that this whole scheme for looking at music is wrong-headed and will yield no answers that get near the truth. When brain research is medically useful, as in treating Parkinson's disease, for example, or aiding in the recovery of stroke victims, the following factors pertain:

- A brain function has gone awry in some organic way.
- The impaired function can be isolated.
- The impaired function can be observed.
- The mechanics of correcting the impaired function are well understood.

When a stroke victim is wheeled into the ER, a brain scan localizes the area of the blood, and the bleeding is stopped using drugs or surgery. Thus all the benefits of treating the brain as a damaged thing are fulfilled. Medical science can look into brain functions with ever-increasing accuracy, allowing surgeons to do their work more finely and leading to drugs whose action is more localized and specific. However, where music is concerned, almost none of the deciding factors are in place:

- No brain function has gone awry.
- The brain functions that produce music are complex and mysteriously connected.
- The actual transformation of noise signals into meaningful music cannot be observed physically.
- There is no explanation for why the higher brain evolved to invent and appreciate music; therefore, no cure exists for people who are totally indifferent to music. It's not a malady.

Is this just a matter of neuroscience being behind the curve? Would a heap of money and more research grants yield better answers? Not if the whole model is fundamentally wrong. The brain somehow produces music out of raw physical data (vibrations of air molecules); everybody agrees on that. A radio also produces music, and yet it would be absurd to say that the two are equal. A radio is a machine working through fixed, predetermined processes. However much it may appear to be similar, the human brain can do anything it wants with musical signals, including tuning them out entirely. Everything depends upon what the mind wants. The brain's mechanisms exist for the mind's use. When a person likes or dislikes a piece of music, the mind makes that decision, not the pleasure-pain centers in the brain. When a composer gets inspired, his mind provides the inspiration, not his neurons. How can one be so sure? The answer would fill a book, but let's divide it into three compartments.

1. Determinism is wrong.

If the brain is hardwired from childhood to hear Chinese music in China, Indian music in India, Japanese music in Japan, and so on, why do all these countries currently have Western-style symphony orchestras, almost entirely filled with native-born musicians, playing Western classical music? You can't call the brain hardwired when connections can be changed at will. Determinism looks good in the schematics of a neurological network, but it breaks down in real life. By analogy, it's as if brain researchers are trying to tell us that house wiring can change from AC to DC current on its own. That's the equivalent of a brain's "deciding" to like Chinese music. Only the mind can create such a shift.

If a dozen interrelated areas of the brain combine to process music, as opposed to processing the sound of a buzz saw or wind in the trees, how does the raw input know where to go beforehand? The auditory center receives all raw data in the same way, along the same channels from the inner ear. Yet the data from a piano go straight into musical processing. This implies that the auditory center already knows which sounds are a buzz saw and which sounds are music, but it doesn't. We see where each signal goes; we don't know why.

Let's roll back the clock to a time when you heard "Let It Be" for the first time. The prefrontal lobes compare new music with a person's expectations from the past. This enables new music to surprise and delight us by defying our expectations. But there are times when new music creates just the opposite reaction in the same listener. You may not be in the mood for jazz one day and yet love it the next. You may be bored by Ella Fitzgerald, only to discover later that you think she's wonderful. In other words, musical response is subject to unpredictable changes. No mechanical system can explain this variability, and reducing it to random neural signals only pushes the problem deeper. The preset chemistry in a neuron can't be expected to produce one response and its exact opposite.

2. Biology isn't enough.

Music exposes why some human behavior makes no sense biologically or in terms of evolution. We love music because we love it, not because our ancestors made more and better babies if their genes carried a response to music. Searching for the evolutionary need for music puts the cart before the horse. Instead of requiring music as a survival mechanism, we enjoy surviving, thanks to music, because our minds delight in it. By any reasonable Darwinian perspective, human hearing should have favored the keenest possible sensitivity, so that our ancestors could hear a lion a hundred yards away instead of ten or twenty. Not getting eaten is a good way to survive. Or, like the Arctic fox, we should be able to hear a mouse move under two feet of snow. More food in winter leads to better survival. But we didn't evolve with that kind of acuity; instead we evolved the totally useless (from the point of view of survival) yet joy-enhancing love of music.

Music is personal, whimsical, and unpredictable. That's not a flaw that science needs to correct or explain. It's part of human nature. On one famous occasion, enemy soldiers walked out of the trenches in World War I to sing Christmas carols together. Which is more human, that behavior, or fighting to the death in a senseless war? Both, in fact. Human nature, like music, is inexplicable in its complexity.

Something new was created spontaneously when "Let It Be" emerged. New styles arise out of sheer inspiration. But let's say that one could build a supercomputer and input every possible musical chord and phrase (there would be more of these than atoms in the universe, by the way), and let's program the computer to develop all possible musical styles. In time it would produce the music of Beethoven purely at random. But that's the very thing that invalidates the computer-brain model, because Beethoven didn't spend a million hours tapping out random combinations until a new style emerged. Instead, a musical genius was born, a single musical mind who listened to the old style, creatively grew beyond it, and changed classical music forever.

3. Your brain isn't listening to the Beatles—you are.

The mind-brain problem, also called the hard problem (see pages 155–157) has proved so impossible because putting the brain first was a mistake. Neurons don't listen to music. We do; therefore, why examine neurons as the key to music, or to any experience? Even the most basic elements of consciousness are absent from the brain. It has no idea that it exists. If you stuck a knife into it, the brain feels no pain. It has no preference for the Beatles or Led Zeppelin. In short, mind cannot be explained by using any object or thing, even the glorious object that is our brain. You wouldn't ask your car radio whether it preferred the Beatles or Led Zeppelin. You wouldn't expect your laptop to cry "Ow" if you stuck a knife into it.

It's time to face facts. There is no physical process that turns air vibrations into music. Inside the brain there is no sound; it is a completely silent environment. "Let It Be," with its qualities of sweetness, religious feeling, pleasure, and all the rest, isn't a product of brain circuitry. It is built from the infinite potential of the mind that gets processed by our nervous system. Music cannot be found in a radio, a piano, a violin, or in a collection of neurons sending chemical and electrical signals to one another.

If we take these facts seriously, the mind assumes a status no machine can duplicate. This status is what we call consciousness. Consciousness cannot be fabricated, which makes it possible to reinvent the universe, not as a place where consciousness somehow got cobbled together on lucky planet Earth two-thirds of the way out from the center of a galaxy called the Milky Way, but as a place where consciousness is everywhere. There are many fence-sitters in physics who will concede that nature acts in mind-like ways, but they cannot swallow the proposition that the universe behaves *exactly* like a mind.

Schrödinger had accepted this impasse almost a century ago, when he declared that it makes no sense to subdivide consciousness. If it exists at all, it exists everywhere, and, we would add, at all times. Therefore, when someone says that consciousness is solely a property of the human brain, they are guilty of special pleading. The brain is doing nothing special that isn't happening throughout the universe. Why is the human mind creative? Because the cosmos is creative. Why did the human mind evolve? Because evolution is built into the fabric of reality itself. Why do our lives have meaning? Because nature proceeds with a drive toward purpose and truth. We promised to answer the "why" questions that crop up everywhere in daily life, and now we hold the key to all of them: cosmic mind drives every event and gives it a purpose.

At this point we've covered nine cosmic mysteries that lead to two conclusions. First, the best answers being offered by science aren't good enough. The "We're almost there" camp wears an optimistic mask, but behind the mask is confusion and deflating confidence. Much less popular is the "We've barely begun to find the answers" camp, but its position holds the overwhelming amount of evidence on its side. It my even be true today that this view is held by a majority of researchers and theorists.

The second conclusion is that reality is trying to tell us something new. It's saying that the cosmos needs to be redefined. All the taboo words rejected by physicalism—*creativity, intelligence, purpose, meaning*—have gained a new lease on life. In fact, we have shown that they are the cornerstone of a conscious universe expressly created for the evolution of the human mind. Reality is the ultimate judge. There is no court of appeals that ranks above it. If reality is pointing the way to a new universe, it would be pointless to resist. Keeping faith with "One day we will know all the answers" does not get us closer to the goal of tackling the nature of reality, here and now.

PART TWO

EMBRACING YOUR COSMIC SELF

THE POWER OF PERSONAL REALITY

What would it take to convince you that you have a cosmic self? Don't settle for a quick or easy reply. Taking on a cosmic self is like taking responsibility for everything we call real. In an epic poem, *Song of Myself*, Walt Whitman proclaimed his universal status with joyful abandon:

> *I celebrate myself, and I sing myself,*
> *And what I assume you shall assume,*
> *For every atom belonging to me as good belongs*
> *to you.*

Rationally speaking, it sounds more than a little preposterous. But readers of poetry didn't take Whitman literally when he declared "I am large: I contain multitudes," and even though no one was better at making ecstasy go viral (as it were), few people were daring enough to follow Whitman's lead when he wrote, "The clock indicates the moment—but what does eternity indicate?" And he had a mind-blowing answer as well. Eternity indicates that human beings are children of the cosmos. Our life is beyond the boundaries of time.

We have thus far exhausted trillions of winters and
summers,
There are trillions ahead, and trillions ahead of them.

Births have brought us richness and variety,
And other births will bring us richness and variety.

This book offers the same answer, not as poetry but as a fact that overturns accepted conventional reality. The cosmic self isn't a pet theory but the most fundamental self anyone possesses. If it didn't exist, neither would the physical world, including all the people and things in it. It is astonishing that a poet singing about himself should match the most far-sighted theories in modern physics, and yet he does:

Do you see O my brothers and sisters?
It is not chaos or death—it is form, union, plan—it is
eternal
life—it is Happiness.

These words apply perfectly to the idea that we live in a conscious universe. Instead of the accepted theory that the properties of mind emerged from churning chaos over billions of years in the distant past, in a human universe, mind has been present at all times and in all places—in fact, *beyond* all times and all places. As an answer to the cosmic mysteries you've been reading about, this one remains standing after all the "reasonable" explanations have been dismantled. *Nothing else makes sense*, do you see it? Too many open problems exist, such as quantum gravity, dark matter, dark energy, and more. Too much of reality is hidden from view. Too many extra dimensions are pure mathematical jiggling to get out of an impasse where theory fails to match reality. The old confidence has collapsed because the building blocks of nature—atoms and subatomic particles—turn out to possess no intrinsic properties without an observer.

In the Sherlock Holmes mysteries, there comes the moment when the great sleuth is about to reveal the hidden solution to a crime and the solution is bizarre and unexpected, such as when a poisonous snake, known as the "speckled band," committed the murder by slithering down a bell cord used to ring for the servants. At such moments Holmes likes to deliver a lesson in deductive reasoning and reminds his trusted sidekick, Dr. Watson, that when every other reasonable explanation has been ruled out, the one that remains, however improbable, must be correct.

To be fair, there's a hole in Sherlock Holmes's lesson in deduction. Faced with a closed-room murder and a handful of suspects, the great detective could relatively quickly exhaust the reasonable solutions. The cosmos, however, is far from being a closed room; it gives almost infinite expanse for newer, more exotic theories, as the last hundred years have proved.

NO ROOM FOR THE MINDLESS

Posing a conscious universe where human life is the whole point cannot simply be another item added to the menu. Unique among competing theories in cosmology, the conscious universe excludes all unconscious ones. They just don't have a reality, and we can't even imagine their reality because it is not there! It is as simple as that.

Being conscious is like being pregnant or dead—either you are or you aren't. No middle ground exists. In our view, the middle ground disappeared once and for all when we showed that the brain doesn't think. The human brain, as a physical thing, can't be the source of mind. By the same logic, the physical universe should be disqualified as the creator of the mind. The universe is huge compared with a human brain, but making a physical mechanism bigger doesn't make it smarter or even capable of thinking in the first place.

Regardless of the shock and outrage among mainstream

scientists, the only way that anything—an atom, brain, or the entire universe—can behave in a mind-like way is to *be* a mind. There is one escape route from this conclusion, however: the so-called clockwork universe of the eighteenth-century Enlightenment. The intellectual trend at the time was to dispense with God as an active participant in how the universe operates from day to day. Yet the processes being observed by scientists—such as the regular order by which the elements fell into place by atomic weight—implied a nonrandom system. The solution was a kind of Solomon's judgment. God was allowed to set the universe in motion with perfect precision, but then he was packed off to heaven while nature's clockwork mechanism kept humming along on its own.

The notion of a clockwork universe seems quaint today, but it was almost the last time that scientists made peace, however uneasily, with consciousness as a serious scientific ingredient when explaining cosmic phenomena. The peace proved temporary. Once God was sent packing, there was never any reason to consider the possibility of a cosmic mind, except in metaphor, as when Einstein declared that he wanted to know how God's mind works, all the rest being details.

Our intention isn't to bring God back, either by marching him in through the front door as creationists do or by sneaking him in through the back way, as happens when mathematics is touted as the ultimate answer to all natural phenomena. Numbers are given a special heaven to live in, as it were. The first philosopher to trace reality to an invisible realm of pure existence was Plato, who held that anything of beauty or truth here on Earth was a shadow of absolute Beauty and Truth in the beyond, reflected, so to speak, in the cave of existence. Today, mathematics occupies the Platonic realm, somehow holding aloof from physical existence in order to arrange it according to perfect mathematical laws.

Being a word for transcendent values, *Platonic* is first cousin to *divine*. There's not much difference between calling the harmony of mathematics a Platonic trait or a gift from God. The problem with shutting God out or letting him in is the same both ways. Consciousness isn't "in" the universe any more than wetness is in water or sweetness in sugar. One doesn't say, "This water is almost right. We just have to add some wetness to it" or "I love this sugar, but it would be even better if you could figure out how to make it sweet." In the same way, consciousness isn't a magic dust you sprinkle on inert atoms to make them capable of thinking. Consciousness has to be there already.

We have seen that mind-like behavior isn't a property of matter. Quite the reverse. When it wants to, the cosmic mind can take on the properties of matter. At the quantum level it can decide to behave like a wave or a particle. When such a choice is made, it's a mental choice, which shouldn't shock us. By definition, choices are mental. We don't say, "My stomach decided to have oatmeal for breakfast." We decide to have oatmeal, not our bodies. The body participates in the choice, of course, because of the mind-body connection. If you're distracted, a rumbling stomach can remind you to eat, just as yawning can remind you to go to bed. Both sides, physical and mental, are allowed to participate.

By turning its back on consciousness, mainstream science made a fateful decision it is slowly beginning to regret. Reality itself seems to demand that ignorance is no longer a valid excuse when it comes to mind and cosmos. The universe didn't become mindless with the stroke of a pen; this was a collective decision made at the outset of modern science. At the time, four hundred to two hundred years ago, a mindless mechanical universe made perfect sense, as we can illustrate through a story everyone learns at school, about Isaac Newton and the apple. The incident is so familiar it would seem to have no hidden dimensions, but it does. It's worth recounting the details as told by Newton to a colleague, William Stuckle. (Spoiler alert: the apple didn't hit Newton on the head.)

> *. . . [W]e went into the garden and drank tea under the shade of some apple trees; only he, and myself. Amidst other discourse, he told me, he was just in the same situation, as when formerly, the notion of gravitation came into his mind. "Why should that apple always descend perpendicularly to the ground," thought he to himself; occasioned by the fall of an apple as he sat in a contemplative mood. "Why should it not go sideways, or upwards, but constantly to the earth's center? Assuredly, the reason is that the earth draws it. There must be a drawing power in matter, and the sum of the drawing power in the matter of the earth must be in the earth's center, not in any side of the earth. Therefore does this apple fall perpendicularly, or toward the center. If matter thus draws matter, it must be in proportion of its quantity. Therefore the apple draws the earth, as well as the earth draws the apple.*

This was a favorite anecdote of Newton's (which scholars believe he in all probability fabricated), although some commentators don't entirely buy the "aha" moment of the falling apple but presume that he had been cogitating about gravity for a while already. In any case, the hidden dimension in the anecdote isn't about what it says but what it doesn't say. Newton and the apple is a prime example of arriving at a truth by excluding everything that doesn't specifically apply. For example, the variety of apple is being ignored, along with the weather, the look of the landscape, Newton's state of health, the clothes he was wearing, and so on. We are so accustomed to excluding all "unscientific" experiences that it has become second nature. We celebrate the fact that the rational mind has this power to focus so sharply and narrowly on nature's mechanics.

On the face of it, reality is inclusive. In fact, it's all-inclusive. Excluding daily experience is an arbitrary mental act. It may yield an astonishing idea like Newton's theory of universal gravitation—particularly brilliant was his insight that the apple's gravitational force draws the earth at the same time as the earth's gravity draws the apple—but exclusion betrays how reality actually works. This didn't particularly bother scientists of the Enlightenment as they dismantled the clockwork universe to discover all its moving parts. But today we live in the "uncertain universe" (one in which there's actually a tiny fractional possibility that an apple can fall sideways or upward, according to quantum probabilities), and the greatest uncertainty of all is reality, slipping through our fingers.

Exclusionism has many successes to its credit, but the human mind is inclusive to begin with. When the waiter places a beautiful chef's creation before you in a restaurant, you don't say, "Give me a moment. I can't decide whether to look, taste, touch, smell, or listen to this food." We take in the whole scene, all the time. (And this happens far beyond the scope of the conscious mind. Under hypnosis subjects can often recall childhood memories with photographic exactness, down to counting the number of steps leading to the attic.) Heeding nature's message to be inclusive conforms to everyday experience.

Newton himself was not a perfect exclusionist. A devout Christian, he believed literally in the chronology of history outlined in the Old Testament. In other words, he was a splitter, allowing natural laws to govern the physical world while bowing to God as ruler of the spiritual world. But being a splitter (or dualist, to use the formal term) was only a stopping point on the journey to total exclusionism once God was removed from the picture entirely in the modern era. In today's landscape, to speak about superstrings or the multiverse involves a conscious decision to exclude all of reality except for a thin mathematical sliver, and even that sliver is just a hypothesis. To reverse course and choose inclusionism implies a seismic shift in how we approach reality. Every time reality is sliced up into data, a piece of the truth is being exchanged for the whole truth—a bad bargain.

Splitting the difference became discredited once God left the building, but he held on for quite a while around the margins. Efforts to weigh the soul as it left the body at the moment of death continued well into the nineteenth century, but in vain. Recently, however, the scientific equivalent of soul research has gained newfound respectability through the concept of panpsychism, which makes mind a property of matter. We think this is a dead end. Panpsychism sounds holistic, which is positive. But it fails to really explain anything; one remains stuck with the mind-like behavior of atoms—that's not an answer, it's the problem needing to be solved. Viewed skeptically, panpsychism looks like the most retro move physics has ever made, jumping back to animism and other aboriginal beliefs that spirit abides in all things.

Even so, the positives of panpsychism are appealing. First, it's a clever ploy to turn mind into a property that all things exhibit. Unlike weighing the soul as it leaves the body at the moment of death, a property doesn't need to have measurable weight and dimension. Nor does it come and go. For example, being either male or female is a property among mammals, but it can't be extracted like drawing blood to see how much each gender weighs or what color it is. Second, panpsychism allows for the universe to act mind-like as its natural behavior instead of an odd peculiarity among quanta. This alone would help make the theory popular—except for a fatal flaw. When you claim that mind is a property of matter, what's equally possible is the exact opposite: matter is a property of mind. One cannot be proved over another. When certain hormones kick in, two people might rush into each other's arms to begin having sex. But just as plausibly, a person can think, I have a little free time. Maybe having sex would be nice, and this thought triggers the hormones. So our behavior, right down to the quantum level, makes simple cause and effect unworkable. To say that matter acts like mind or that mind acts like matter won't do, either way. Otherwise, we wind up saying weird things like, "The wetness of water is what made people want to swim." A mere property isn't a cause.

Human experience is the last thing anyone should want to exclude when explaining the cosmos. Let's see if we can acquire a vocabulary of inclusion. Reality is all-inclusive, without a doubt, and almost miraculously, human beings can embrace an infinite variety of what reality has to offer. Where is the switching mechanism that decides to gaze at a gorgeous sunset while ignoring the texture of the ground beneath your feet, or indulging in the touch of a beloved while totally shutting out what the furniture in the room looks like? We do these things so automatically that we take them for granted. The critical issue is what it means to experience the world.

The answer is that we experience the world through choice. There is no given world. If Newton's apple was anything like the ones sold at the supermarket, it was red, sweet, crunchy, slightly grainy in texture, and within a certain range of weight. None of these properties exists in nature. They are perceptions of the human mind. The apple doesn't need to be reinvented every time you encounter it. Once your perception has decided that apples taste like apples rather than pears or avocados, they stay that way in your mental setup. As we saw earlier, reality is filtered through the brain and its built-in limitations (recall the revolutionary thinking about this by Arthur Korzybski, discussed on pages 139–140). But the brain's imperfection doesn't negate a simple fact: Everything we perceive is a mental creation, accumulated over millions of years of evolution.

It sounds strange to say that we chose for apples to be sweet, because that happened long, long ago. Once sweetness became part of our perception, it was expressed physically in our taste buds, which in turn are encoded in our genes. A separate apparatus for liking or disliking sweetness is encoded in our brains. But change is always possible. If you are ill enough with the flu that nothing tastes good, for example, an apple's sweetness can be totally erased by your perceptions. As conscious beings, we still aren't universal perceivers. Our eyes can't see objects in pitch-black darkness. If the human brain could detect ultrasound frequencies and infrared light—traits found elsewhere in nature, among bats, sharks, reptiles, and so on—those abilities would get translated into how our brain functions. Yet we can go beyond our limited hardwiring by developing instruments for detecting frequencies of lights and sound where our senses leave off—in that sense, we've turned ourselves into potential universal perceivers after all. As choice-makers go, we seem to be the champions in nature.

There seem to be many things we can't choose to change, such as gravity, the hardness of rocks, and the solidity of a brick wall. Some distinctions need to be made therefore. Our perceptions come in three types:

Perceptions we can't change.

Perceptions we can change.

Perceptions that sit on the borderline, being sometimes changeable and sometimes not.

In your personal reality, all three kinds mix and match. If you don't like the color of the shirt you're wearing, you can change it—that qualifies as a changeable perception. If you can't walk through walls, that falls among the unchangeable perceptions. One could continue with hundreds of examples from each category. The spice of life comes from the perceptions we change, while the solid security of life comes from the ones we can't change. If you could decide not to obey the law of gravity on Mondays, a world of chaos would ensue, beginning with your body vanishing in a misty cloud of atoms.

But what's truly fascinating is the third category: the perceptions we can sometimes change and sometimes cannot. This is where quantum theory made our participation in nature more puzzling and more enticing at the same time. It created a shadow zone where particles and people both can make decisions. Being passively present without participating was no longer an option. Every perception is an act of participation in reality. If you perceive another person as the love of your life, your actions will lead you into areas of reality unknown before that perception. Every day our actions exist on the cutting edge of evolution, the frontier where the mind is caught between caution and curiosity. The most obvious example is miracles. Who wouldn't love to believe that a human being once walked on water, that faith can cure cancer overnight, that the dead are in communication with the living? The controversy over miracles isn't in whether they can occur, however, but in what category they belong to. A miracle is only available if it fits into the third category, things that sometimes occur and sometimes don't. Of course you can always practice total exclusion (the fixed attitude of atheists and skeptics) or inclusion (the fixed attitude of the religiously devout).

And if you have no fixed attitude? Then you belong in the company of visionary quantum pioneer Wolfgang Pauli, who said, "It is my personal opinion that in the science of the future reality will neither be 'psychic' nor 'physical' but somehow both and somehow neither." By using a word that science shuns—*psychic*—Pauli was pointing to a kind of ultimate mystery. The vast physical mechanism we call the universe is on dual control, obeying natural laws and thoughts at the same time. This is the basic reason we presently occupy an uncertain universe. But Pauli pointed the way to a solution when he predicted that reality's amalgam of mind and matter would be both and neither at the same time. This sounds like a paradox, so we'll unravel it to reveal why Pauli was simply stating an undeniable truth.

QUALIA: REALITY IS UP FOR GRABS

Let's bring this discussion to the personal level. Which parts of your own reality can you change, using only your mind, that actually make a difference? The answer requires a new term in our kitbag: *qualia*. The concept is tremendously important, even though the average person has never heard of it. With qualia you can change your perceptions—or not. With qualia you can alter reality—or not. Qualia refers to how we experience life rather than how we measure it. The word *qualia*, which is Latin for "qualities," is a tag for a world that is as far-reaching as quantum physics but points in the opposite direction, away from physical objects and toward subjective experience. Whereas quanta are "packets" of energy, qualia are the everyday qualities of existence—light, sound, color, shape, texture—whose revolutionary implications we've already begun to describe.

You experience the world right this minute as qualia. It's the glue that holds the five senses together. The scent of a rose is a qualia (we use the same word in singular and plural), as are the velvety texture of its petals, their colors and hue, shadows and folds. Looking at everyday experience through the perspective of the brain, psychiatrist and neural theorist Daniel Siegel's model for reality "in here" is SIFT: sensation, image, feeling, thought. No matter what's happening to you right this minute, your brain is registering either a sensation (I'm hot, this room is stuffy, the bedsheets are soft); an image (the sunset is brilliant, I see my grandmother's face in my mind's eye, my keys are on the dining room table); a feeling (I'm pretty happy, losing my job makes me worried, I love my kids); or a thought (I'm planning a vacation, I just read an interesting article, I wonder what's for dinner).

Qualia are everywhere. Nothing can happen without them, which means that if you participate in reality using a human brain, your world consists of qualia. If there is a reality that exists outside what we perceive, it is inconceivable, literally. Once you subtract everything you can sense, imagine, feel, or think about, there's nothing left.

Here's the kicker. Because qualia are *subjective*, they directly attack the objectivity of modern science. Moreover, because experience is *meaningful*, qualia attack the model of random, meaningless nature. Yet even more is at stake.

As its most revolutionary claim, qualia science declares that *only* subjective experience is reliable. At first glance this statement seems preposterous, especially for a scientist. Subjectivity is notorious for being unreliable. Are people given the right to say, "I don't like gravity. Take it away" like a restaurant customer who doesn't like the looks of his entrée? No, because as we saw, some things can't be changed simply by wishing them to change. Yet the argument about unreliability doesn't hold water. It is only plausible if your yardstick is measurement. If a stranger asks for directions, and person A tells him to go west one mile while person B tells him to go east two miles, a map settles which one is right.

But measurement is a red herring. Einstein proved once and for all that nothing—which means absolutely nothing—is immune to relativity, and relativity is all about perception. If you are riding in a spaceship lifting from Earth, your body is exposed to many Gs of gravitational force. An astronaut feels incredibly heavy during liftoff, and his perception is the real thing. Acceleration, according to Einstein, is the same as "real" gravity. Likewise, the color blue doesn't exist without an eye that responds to light like the human eye. If a Martian landed and said with admiration, "The sky on Earth is grimmick," a human being would have no way to understand what is meant, because "grimmick" isn't a color in our reality, and we don't even know if grimmick is a color.

Qualia are the true building blocks of reality. You can lead your whole life without taking a scientific measurement, but a scientist cannot do anything without sight, sound, touch, taste, and smell. If you love the smell of boiled cabbage, while someone else hates it, this doesn't prove that subjectivity is unreliable. It proves that we have infinite creative freedom in the playground of qualia.

So-called objective measurements are just isolated snapshots, a snatched glimpse of the actual fluidity of experience. These snapshots are at once true and false. Imagine that you are a worried father of a wayward teenage daughter, and you've hired a private detective to follow her. After a week he brings you a batch of photos. One shows your daughter trying on shoes, while others show her flashing a fake ID at a bar, sneaking a smoke in an alley, and texting to a girlfriend at the movies. Each snapshot is true, but as a composite, they capture nothing essential about your daughter except that this composite has many facets that are loosely connected. A set of snapshots taken the following week, which show her visiting a sick friend at the hospital and volunteering at an animal shelter, contradicts the pattern suggested by the first set. Physics finds itself in the same position, with the exception that it must fit together thousands of isolated observations, and the most basic ones, focusing on subatomic particles, last only a few thousandths of a second.

By contrast, qualia are constant and continuously connected. If you replace the snapshots of nature's details with an endless movie, the universe is actually a mirror of the human nervous system. Physicist Freeman Dyson supports this conclusion: "Life may have succeeded against all odds in molding a universe to its purposes."

Behind the mask of a cosmic machine whose parts can be calculated and tinkered with, the universe is humanized. There is no other way it can exist, in fact, since nothing "out there" can be experienced except in our own consciousness. We are following the trail pioneered by physicist David Bohm, among others, when he wrote, "In some sense man is a microcosm of the universe; therefore what man is, is a clue to the universe."

BUT . . .

When physicalists get backed into a corner, their defensive tactics aren't subtle. To discredit qualia, examples like the following are frequently used: "Forget your metaphysics. Reality is a given.

If you get hit by a bus, your whole theory goes out the window. You'll be just as dead as anyone else." Getting squashed in an encounter with a bus seems like a convincing outcome to our common sense, and for the bus you can substitute a car, train, or brick wall. But physicalism can't explain why a bus, train, or brick wall is hard to begin with, given that all matter is more than 99.9999 percent empty space. The standard answer, that hardness results from the opposition of electromagnetic charges, is like handing you the chemical formula for sucrose to explain why sugar is sweet.

Second, qualia aren't free-floating and temporary. Some qualities, like the wetness of water and the hardness of a brick wall, are set in place. They form structures that are just as real as the formula for sucrose. The big advantage is that sweetness is an actual experience, whereas the formula for sucrose is only the map of an experience, and you can't get from the map to real life.

The conscious universe embraces change, nonchange, and the state of potential change. This is another reason, and one of the most important, why the cosmos feels completely humanized once you open up to the possibility. We saw that there are perceptions that we can change, those we cannot change, and others that we may or may not be able to change. These perceptions are the world created by the building blocks of qualia. The fact that a human body will be crushed by a moving bus belongs to the setup that isn't changeable. It says nothing about how the setup was created in the first place.

If we knew how the setup was created—and is still being created—we'd unlock the secret of how reality evolves. Our cave-dwelling ancestors had already evolved the higher brains (the cerebral cortex) that are barely different from the cerebral cortex of Einstein or Mozart. Yet in hunter-gatherer society there was no need for an Einstein and a Mozart. No survival purpose would have been fulfilled by them. Instead, for reasons that remain a mystery, the cosmic mind fashioned a brain machinery capable of infinite adaptation. While early *Homo sapiens* was preoccupied by the technology of making flint arrowheads and stitching hides together with animal sinew, the higher brain was already outfitted for the future, for Mozart sonatas and quantum mechanics.

So who knows what our own brains are already outfitted for that will come into play a thousand or ten thousand years in the future? It's quite miraculous that evolution is able to see beyond the next horizon in this way. For there is no doubt that other higher primates, like the chimpanzee, also created primitive tools, and yet they hit an evolutionary wall somewhere along the line. A chimp's ability to go beyond its present abilities is severely limited. Ours isn't. Human history is filled with untold horrors of war and violence, and yet our brains are also set up for Buddhist meditation, Quaker pacifism, and mystical ecstasy.

In short, the human universe depends on seeing beyond our current abilities, where we feel trapped by the physical world and hemmed in by its rules. Cosmic mind isn't done with us. A powerful evolutionary force has propelled the human cortex to unparalleled heights at unbelievable speed. The rise of the higher brain took less than thirty thousand to forty thousand years, a blip on the screen of evolutionary time. To discover where the evolutionary tidal wave is headed, we need only explore one of the most amazing human traits, shared so far as we know by no other living creature. We are aware of being aware. The next horizon, it turns out, is inside us, and if we want to take the next leap forward in our evolution, the only map is the one we create for ourselves, in our own consciousness.

WHERE YOU REALLY CAME FROM

Being connected to the cosmic mind is built into your nervous system. You were born to see light and hear sounds. Those abilities can also be traced to your nervous system. Specific areas will light up when music vibrates your eardrum and fireworks glow in your retina. But the cosmic mind has no specific location in the brain. How do we know that the cosmic connection is real, or that it is doing anything for us? A skeptic might point out that the lives of countless millions of people contain misery, poverty, and violence. Even the most fortunate lifetime will be visited with accident and disasters along the way. The skeptic will ask, of what earthly use is your so-called cosmic connection if it can't relieve the difficulties of everyday existence?

For our answer, we have to look deeper into the setup of mind, both individual and cosmic. We mentioned that some things can be changed while others cannot, and in a third category belong things that may or may not be changeable. In fatalistic societies, such as medieval Christian Europe, God was thought to be so powerful that the individual had little room to improve his lot in life. The present era, by contrast, is filled with aspirations. People seek not just self-improvement but total transformation, which is why the notion of a conscious universe is taking hold right now and with such force. Such a universe is constructed to promote the expansion of consciousness in the individual. On that basis alone we can talk about change and how to achieve it.

Think of the world you are familiar with—a world of family, friends, work, politics, leisure time, and so on—as a self-enclosed system. Within this system the parts fit together and hum along, giving little hint that there's a larger reality outside the box. If you are unaware of this larger reality, the potential for change is limited by what is allowed in your world. You can't change what you aren't aware of. Therefore, the conscious universe might as well not exist, because it has no effect on your daily life. Skepticism would be a normal, natural response if someone told you that you were connected at every second of your life to cosmic mind.

Now consider the opposite extreme, an existence marked by total detachment from worldly things. Someone who has arrived at complete detachment—a Yogi or Zen Buddhist monk, let's say—has no allegiance to how events work out. Good and bad, pain and pleasure no longer generate the response of wanting more of the good and less of the bad, more of the pleasurable and less of the painful. The human nervous system is infinitely flexible, and any of us could embrace such an existence, with its pure, peaceful stasis, if we had a mind to. We would be free of any system, but at a cost. We would renounce most of the things ordinary people passionately pursue, because in our detachment, change is meaningless; to gain or to lose is the same. As spiritual as this may sound, to renounce the world may be just as divorced from cosmic mind as living a totally worldly life.

Which leaves the third option, where some change is possible and some isn't. We can call this the evolutionary choice, because your life is driven to seek more awareness and to enjoy the fruits of awareness through love, truth, beauty, and creativity. But at the same time you embrace the peaceful, centered detachment that underlies all of existence. This third option—change in the midst of nonchange—is the one we favor, because it makes full use of the connection with cosmic mind. On the one hand there is immense dynamism and change; on the other there is the reality of pure awareness, the silent source from which all of creation springs.

Once you grasp what the options are, it is clear that terms like *objective* and *subjective* no longer apply. Outer life and inner life move as one. Daily activity is still individual—you are the specific person who wakes up, starts the car, and goes to work—but the consciousness that creates reality is universal. Intriguing as this sounds, we still have to prove that the connection to cosmic mind is real, workable, and an improvement over life being lived without such a connection. If you came from the realm of pure awareness, not simply from your mother's womb, understanding that can cause true transformation of the kind so many people seek and crave.

"MY" MIND OR COSMIC MIND?

Abstractions are always a danger, and it may seem, even this late in the book, that cosmic mind is too abstract a concept to be real or practical. Let's say you are planning a vacation and can't decide between the mountains or the seashore. After searching for hotels, you find a great deal on a hotel in Miami Beach, and that tips the balance. Now, did this whole process take place in cosmic mind? We are used to phrases like "I've made up my mind" and "I can see it in my mind's eye." They imply that each person possesses their own mind, so it's "my" vacation, search for hotels, and decision to go to the beach.

But this is the very illusion that makes reality "out there" separate from us. In a dualist setup, "my" mind is different from cosmic mind. It's much smaller, for one thing, and its viewpoint is limited to the experiences I've had since birth. Yet if we abandon the illusion of separation, there is no need to choose either/or. The mind *feels* personal and at the same time it *is* cosmic.

Imagine that you are a single electron flickering in and out of the quantum vacuum. As a single particle you feel like "me," an individual. But in reality you are an activity of the quantum field, and in your guise as a wave instead of a particle, you exist everywhere. In our daily lives we are accustomed to feel like individuals while overlooking that at another level, every person is an activity of the universe. What is true for an electron is true for structures like the human body that are constructed from electrons (and other elementary particles).

When you live in separation, ignoring your holistic self, life resembles presliced bread. The urge to divide and subdivide allowed science to claim, quite falsely, that objectivity and subjectivity were entirely different, with objectivity being the superior part. But the quantum era abolished this neat division, and reality started to lead in a new direction—the very things we've covered in the preceding chapters.

But can reality ever be seen directly, as a whole, with no divisions or separations? It sounds like a spiritual quest, which a prior age would call union with God or atman or satori. Reaching beyond separation was motivated by the desire to commune with spirit and at the same time to escape earthly suffering. Now the urge is different, focused much more on higher consciousness and fulfilling one's potential. Finding a new motivation is as important, however, as trying to understand where we came from, because only certain knowledge can assure us that the cosmic mind is our source. Once we are sure of that, birth and death are seen in a very different light, under the aspect of eternity.

The habit of slicing reality into neat manageable pieces is hard to give up, largely because a holistic approach seems literally impossible. At least this is what everyday experience seems to imply. How do you look at the whole human body instead of cells, tissues, and organs? How do you look at the cosmos beyond space, time, matter, and energy? We shouldn't exaggerate the difficulties of being a whole person. If we refer to everyday life, the body isn't experienced as cells, tissues, and organs. Rather, it is experienced in different *states*. Being awake is a different state from dreaming and sleeping. Feeling ill is a different state from feeling well. As we've seen, quantum mechanics works in similar fashion. A wave is a different state from a particle.

Likewise, mind and matter are considered so different from each other because that's our habit of thinking, but really, mind and matter are different states of the same thing: the field of consciousness. You can follow them as they morph from one to another by looking at the brain, where mental events create brain chemicals in one seamless motion. Thus, if a near collision on the highway causes you to be frightened, that mental event translates into molecules of adrenaline, which in turn translate into physical changes such as dry mouth, pounding heartbeat, and tight muscles. When you notice these changes, you are back in the realm of the mind. Likewise, all kinds of signals travel on a journey of transformation from physical to mental that has no definite end point. Life is transformation itself.

What happens in our bodies is also happening in the universe, where any event belongs to the constant transformation of consciousness into either mind or matter. But such a statement explains nothing until we know what consciousness is. If "my" mind, "my" body, the billions of galaxies in outer space, and cosmic mind can all be reduced to states of consciousness, it behooves us to settle once and for all what consciousness actually is. Otherwise, we're just pretending that chalk and cheese are the same thing when it's obvious that they aren't.

First of all, consciousness can have many states, so it doesn't appear as one thing even though it is. If you are dreaming of a beach in Jamaica, you could be having a so-called lucid dream where all five senses are engaged. You can feel the warm sand under your feet and smell the scent of tropical flowers carried on a sea breeze. But the instant you wake up from your dream, you recognize that you were simply in a special state.

Knowing what state you are in is the key to wholeness. Imagine two sportscar drivers. One driver's car has five gears, and he is skillful at switching from one to the other. The second driver has five cars, each equipped with one gear. For him, driving isn't holistic and unified, because it depends on which car he chooses to drive, and each of them is confined to one gear only.

The challenge is to navigate our way through a cosmos where every gear (space, time, matter, and energy, plus other physical properties such as electric charge, magnetic field, etc.) is interchangeable. The whole thing could dissolve into quantum soup if there wasn't an organizer whose viewpoint took in everything, and cosmic mind serves as just such an organizer. Time, space, matter, and energy are managed from the same gearbox, and the driver (consciousness) selects which state he wants to be in. Reality consists of shifting, interchangeable states that emanate from one source: consciousness.

GIVING THE UNIVERSE ITS EVICTION NOTICE

It is alluring to think that we exist in a living universe. By definition, if the cosmos has a mind, it must also be alive. But whether you call it a conscious universe, a living universe, or (as we have done) a human universe, problems creep in. One problem is practical. How do you live in a conscious universe? Would you shop for groceries, go to birthday parties, and gossip around the water cooler any differently? The answer is yes. A conscious universe is totally transformed from the uncertain universe we now occupy, and the transformation runs so deep that it calls all behavior into question. As Peter Wilberg, one of the most astute and gifted qualia theorists, has explained, we don't see because we have eyes. Eyes are physical organs that evolved to serve the mind's desire to see. Mind comes first. It reaches out to experience reality through qualia, which embrace the five senses, along with sensations, images, feelings, and thoughts in the mind.

The spiritual rebirth that every saint, sage, and mystic has promised depends on a new reality, which means a new universe. Or, rather, a new way to look at the universe already there. There's a huge roadblock to such dreams of renewal, and it's the second problem we face when approaching reality as a whole. *The limited mind can't do it.* It cannot think its way to renewal, imagine its way, feel, see, or touch what transformation would be like. The linkage between the uncertain universe and the mind that created it is as strong as iron. In other words, if the mind is trapped in its own perceptions, how can the same mind free itself? We seem to face another snake-biting-its-tail predicament.

Here a new term will be useful: *monism*. Derived from the Greek *monos*, which means one, alone, or unique, monism is the alternative to dualism. Oneness is the basic trait of reality, not separation. In some forms of monism, everything in existence is part of the body of God. Other forms of monism view the universe as made of only one substance. Physicalists, who believe that everything can be traced to a material source, stand for one school of monism. Einstein's search for the unified field, the Holy Grail of science, is monism. The rival school, which believes that everything is made of mind, used to be called idealism, but this term became so discredited that we'll use the term *consciousness* instead.

Imagine that you aren't allowed to cast a ballot in the next presidential election until you declare which monism you belong to, physicalist or consciousness (which we've also tagged "matter first" versus "mind first"). How would you choose? Everyone's mind is hopelessly conditioned, burdened by all the old choices it has made, and these old choices, going back to the first hours of infancy, turn out to be self-centered. There's an urge in the development of children that says, "I have to be me"—in other words, a freestanding individual. But projecting this urge onto the cosmos causes dualism to run amok. It turns the usefulness of a separate self into a law of nature, which isn't so.

In everyday life, dualism falls into categories everyone is familiar with:

What you like versus what you dislike

What causes pleasure versus what causes pain

What you want to do versus what you don't want to do

People you like versus people you dislike

In short, it's an either/or world built of opposites. The opposite of before is after, the opposite of near is far, the opposite of here is there. But these paired opposites are not actually real. They are mind-made. So if you want to get real, all the mind-made stuff must be discarded. At the most mundane level, if you judge people by the color of their skin, you cannot know who they really are until the concept of skin color no longer has any bearing on the case. It can take many decades to heal this one symptom of dualism; one can only imagine how hard it is to throw dualism out completely. The process goes far beyond personal values; in essence, it means giving the universe an eviction notice. Because the subatomic particle has no fixed properties, neither do the things made of particles. If you take this seriously, all physical objects have to be evicted, from quarks to galaxies.

Objects cannot exist without space, so when objects are sent packing, space has to hit the road, and since space is in a relativistic relationship with time, according to Einstein, time doesn't get to hang around either. The present state of physics, in some quarters at least, has gotten this far. The perspective in which matter, energy, other physical quantities, time, and space are deprived of absolute fixed reality can be called weak dualism because, as heroic as it is to depose the material universe, we haven't arrived at wholeness yet. Once it occurs to the mind that the material universe was mind-made all along, there's not much reason for the mind to trust itself. Some scientists contemplate the mind's ability to create qualia and conclude, wrongly, that there is no meaning to anything, that the whole cosmos is pointless.

This loss of confidence can be productive, however, if it motivates the next stage of the journey to wholeness. In order to stop believing in self-created illusions, the conditioned mind is also served an eviction notice, this time *by itself!* Only then can cosmic mind enter as a replacement. It's rather like a cardiologist performing a heart transplant on himself, only trickier. Rupert Spira, a brilliant spiritual teacher, calls this the acceptance that some things are not mental events. Death is one example. The mind would like to survive death, Spira jokes, so that it can come back and say what the experience was like.

In its nature the mind isn't an activity at all, but something else. Just as a lake isn't essentially the waves that ripple across its surface, the mind isn't the activity of thinking, feeling, sensing, or imagining. A lake is a still body of water; the mind is awareness without waves. This is the unchanging backdrop of everything that comes and goes. There are no longer any mental events to cling to, and steadily over time, silent mind becomes like home, like a resting place where you truly belong. The good news is that lacking mental events, the mind doesn't die. Instead, it does exactly what was required all along: it changes state. In this case, the change is from constant thinking, wishing, fearing, desiring, and remembering (i.e., the experience of separation) to a state in which one is simply conscious, aware, and awake (i.e., the experience of wholeness). The choice to make this shift is ours. Being infinitely flexible, reality permits the experience of separation to be totally convincing and the experience of wholeness to be totally convincing. But the two states certainly feel different. Here are some examples of how separation is experienced.

What Separation Feels Like

You see yourself as an isolated individual.

You listen to the demands of your ego and place "I, me, and mine" ahead of other people's.

You are powerless in the face of mighty natural forces.

The basics of survival require work, struggle, and worry.

You long to join with another person in order to solve the problem of loneliness.

The constant cycle of pleasure and pain is inescapable.

You may find yourself prey to mental states beyond your control, such as depression, anxiety, hostility, and envy.

The external world dominates over the inner world—hard reality is inescapable.

When you ask other people if they are in the same state of separation as yourself, it turns out that they are. Since everyone is in the same soup, it becomes accepted as reality. What's fascinating about this list isn't how much misery it outlines, although there is more than enough of that. The fascination lies in the linkage between everything on the list and the behavior of the universe. As several quantum pioneers pointed out, the universe displays whatever the experimenter is looking for.

By contrast, here is what it feels like after the illusion of separation has fallen away.

What Being Real Feels Like

You are not in the universe. The universe is in you.

"In here" and "out there" are mirror reflections of each other.

Consciousness is continuous and present in everything. It is the one reality.

All the separate activities in the universe are actually one activity.

Reality isn't just fine-tuned. It is perfectly tuned.

Your purpose is to align yourself with the creativity of the cosmos.

The next thing you feel like doing is the best thing you can do.

Existence feels free, open, and without obstacles.

Mind and ego still exist, but they get a lot more time off.

Knowing who you really are, you set off to explore unknown possibilities.

Probably the first point, "The universe is in you," seems the most baffling. As a declarative statement of physical fact, it borders on the absurd, since billions of galaxies obviously cannot be confined inside a human being. Where would they be? Inside the skull? Clearly not. But "the universe is in you" comes at the end of a journey; it's not an isolated idea. Along the journey we saw that every experience occurs as qualia—in other words, qualities like color, taste, and sound. Since qualia occur in consciousness, they aren't limited by physical dimensions. No one can boast, "Blue is a much bigger color for me than it is for you," or "I keep my vocabulary in a locker in Los Angeles because I go there so often."

Because qualia have no dimension—they aren't short or tall, fast or slow, and so forth—it is entirely possible for a cold virus to occupy the same "space" as a billion galaxies, when we are speaking of mental space. Blue has no specific home except in consciousness. You can either call it to mind or leave it be. The same is true of your vocabulary. You can call upon the word *giraffe*, while letting the rest of your vocabulary exist in mental space, which is everywhere and nowhere. The brain is made of qualia. It has the texture of stiff oatmeal, contains miniature watery lakes, and exudes various secretions. All of these qualia also occupy the same "space" as a cold virus and a billion galaxies. They are all in consciousness. What we commonly refer to as "outer space" is just another qualia. You might protest, "Look here, my brain is inside my skull, and there's no getting around this fact." But imagine the face of someone you love. The brain produces the image in such a way that it isn't inside its tissues—no matter how hard you search, you won't find any images in the brain.

So it must be true that the brain serves one function: it gives access to the mental "space" where all concepts, experiences, memories, images, all qualia reside. A radio gives access to a hundred-piece symphony orchestra without anybody tearing it apart to see the hundred musicians hiding inside. Yet neuroscientists find it hard to stop doing the same thing. They want the brain to be the place where consciousness lives when in reality the brain is only the *doorway* to where consciousness lives. Why did consciousness need such a doorway? For the same reason that getting hit by a bus hurts or even kills you. Consciousness has the innate habit of creating things, events, experiences. This is its natural behavior. Max Planck had this in mind when he said, as we've quoted several times, "I regard consciousness as fundamental. We cannot get behind consciousness." Reality doesn't have to give an explanation for how it behaves, because it has nothing to answer to but itself.

THE MIND AS CREATOR

This leads to a new stage of the journey in which your mind sees quite clearly that it is the maker of your personal reality and has functioned as the maker all along. In itself, this isn't a profound insight. Anyone who has fallen in love only to discover, months or years later, that their beloved is an ordinary person, knows the power of mind-made reality. The real insight is to see that mind-making uses no bricks or mortar, not even the finest matter, energy, time, and space but only one thing: concepts. Take the concept of "I," the separate self. The instant the mind thinks "I," which is the root of all separation, the entire universe falls into line as a world apart from "I."

The whole setup would be hollow and boring if "I" saw through the illusion, so it comes up with a multitude of experiences that keep separation going. For many people, science proves that the illusion "works." They are as sure of the moon and the stars as of anything in existence. It took imagination, skill, and ingenuity to send the Hubble telescope into space and investigate still farther "out there." That's a considerable upgrade of the illusion from squinting at the stars with the naked eye. But providing yourself with a better view of the illusion doesn't make it real. By the same token, in a dream where the sun is shining, does the dream become real if two suns are shining, or a dozen, a thousand, a million?

Having seen that it builds reality out of nothing, the mind may pause to marvel at how amazingly convincing the state of separation is. This is what we meant about reality being infinitely flexible, allowing separation to flourish for as long as it is convincing. You could spend your entire waking life searching for new orchids, finer cuisine, more beautiful women—whatever qualia you desire. Since every experience consists of qualia, you can even tell yourself, "Relax, this is all there is." To be honest, there's a faint sadness when you see through the illusion. To know that orchids, cuisine, and the beauty of women are all mind-made creates a feeling of hollowness—for a while.

The mind decides that there must be a better world elsewhere, and this new challenge defeats the sense of sadness. Like a painter throwing his palette into the garbage can, the mind decides to get rid of imaginary concepts. It's a very bold decision, because even the universe is one massive concept. *Any* concept leads to the state of separation. Only reality is exempt. It is not mind-made; therefore, reality is inconceivable.

To realize this fact—which means to experience it personally, not just as a nifty idea—creates a great pause. *Oh my God*, you realize, *I am never going to grasp what is truly real. It lies beyond my mind, my senses, my imagination.* Now what? The great pause doesn't have to be spiritual, although it was for Gautama sitting under the bodhi tree or Jesus on the cross, saying, "It is finished." The great pause can be found in the words of a scientist, including Heisenberg and Schrödinger, who suddenly sees, quite clearly, that there is only one reality, not two. There is no inner and outer, no me and you, no mind and matter, each half jealously guarding its own marked-off territory. This realization is like a pause because the mind has stopped conceiving of reality and now starts *living* it.

DUELING MONISTS AT THE O.K. CORRAL

The argument for a conscious universe has been swirling for more than a decade among cosmologists and the conferences they attend. But you don't read headlines such as "Universe Turns Topsy-Turvy." The number of theorists who started out as physicalists, only to realize that consciousness is everything, isn't zero, but it's not far from zero, either. In some horror movies the hero has done all the right things—shot the vampire in the heart with a silver bullet, warded off Dracula with a cross, or exposed him to the withering light of day—and still the creature keeps coming back. Physicalism keeps coming back, for the most part, because of a mental habit we discussed almost at the start: naïve realism. "If you get hit by a bus, you're dead" refutes all objections, end of story.

A more sophisticated objection can be called "the case of the dueling monisms." Proponents concede that reality is indeed one thing, but that one thing is physical, not mental. Here's how the argument might go.

Physical monist: "You say that the universe is mind-made. Within your monism, mind turns into matter, but you don't say how. The brain isn't where the mind lives, according to you, but if you cut off someone's head, there's not much mind left. So the only thing your consciousness model has going for it is that you happen to believe in it.

"Well, surprise, we have a monism, too. In it, there's a physical process behind everything. We can measure these processes. They fit beautifully with mathematical predictions. Scanning the brain, we can observe the mind at work. Our monism is just as consistent as yours, and it's supported by a mountain of evidence."

You've now read dozens of ways to refute this argument, but it's clear that mere refutation isn't good enough. Technology is science's ace in the hole, and there's an implicit threat that if we give up the physicalist approach, the world will slide backward into primitive times. Technology will be stalled by airheaded mystics and philosophers. People love their iPhones and flat-screen televisions, all the technology that the physicalist approach has created. Would anyone risk losing all of that? This isn't a veiled threat. In repeated interviews, popular planetary scientist Neil deGrasse Tyson has warned that philosophy is worse than useless compared with science. Two examples—

1. "My concern here is that the philosophers believe they are actually asking deep questions about nature. And to the scientist it's what are you doing? Why are you concerning yourself with the meaning of meaning?"

2. "[D]on't derail yourself on questions that you think are important because philosophy class tells you this. The scien-

tist says look, I got all this world of unknown out there. I'm moving on, I'm leaving you behind. You can't even cross the street because you are distracted by what you are sure are deep questions."

The confidence behind these assertions ignores the fact that the deep questions deGrasse Tyson disdains were brought up by the greatest quantum physicists of the last century. Let's set that aside. We can take a different tack, which is to show that consciousness offers a better life than technology. It opens up a future in which the planet can be saved from potential destruction. It puts the individual at the switch where choices change personal reality. At the same time, "all this world of unknown" will be supplied with answers that only consciousness can supply. If we can accomplish all of these things in our concluding chapter, the duel of the monists at the O.K. Corral will be over. And when it is, everyone still gets to hold on to their iPhones.

HOME FREE

Hero worship gets you only so far. We've been holding up the first generation of quantum pioneers as the Greatest Generation, not as warriors but as seers. Instead of storming the beaches of Normandy, they stormed the beaches of time and space and, ultimately, the mainland of reality. But, as one physics professor at the California Institute of Technology retorted when he heard Einstein's name used reverently, "Any Caltech grad student in theoretical physics today knows more than Einstein." A sizable proportion of working physicists would agree. Einstein, Heisenberg, Bohr, Pauli, and Schrödinger would be eating our dust, so far behind was their thinking.

None of the quantum pioneers, for example, possessed our knowledge of the big bang, and no amount of hero worship can get around that fact. The cosmos behaves today exactly the way it should behave if the big bang occurred 13.7 billion years ago, and until it behaves differently, the hypothesis of the big bang is king of the hill.

Turning to a conscious universe would make the big bang an incidental concept. The new king of the hill would be qualia, the qualities created in consciousness. A flickering candle gives off heat and light, and so did the big bang. But without the human experience of heat and light, creation as we know it couldn't exist. (Note how baffling "dark" energy and matter are. We are still searching for the qualia that match them.) That's why qualia come first and even a tremendous event like the big bang is secondary. What keeps the physical universe intact *is* qualia.

If qualia became an unquestioned part of our understanding, would it revolutionize everyday life, which is our position, or would people shrug their shoulders and carry on as usual? A conscious universe will only gain traction if we can humanize it. Otherwise, the status quo, an uncertain universe, will continue. As a concept, the uncertain universe has proved a remote, random, hostile environment that has no place for human beings except as a cosmic accident. Instead of being winners in the cosmic casino, we may be cosmic dodo birds waiting for extinction. It's not as if the multiverse needs us. A trillion trillion rolls of the dice should bring about a new universe suitable for our kind once more.

Our hero worship was justified, and we are hardly alone in quoting Planck, Einstein, Heisenberg, Bohr, Pauli, Schrödinger, and others as modern prophets. It's quite usual, in fact, to haul them out if you want scientific backing for believing in higher consciousness. The spiritual side of the quantum pioneers, though an embarrassment to mainstream science, is a beacon for seekers. The problem is that our heroes didn't follow up on their tremendous insights into consciousness. Their working lives actually did more to create the uncertain universe than anything else. Perhaps it could not have been otherwise. After all, they were trying to build a radically different way to study the physical cosmos, not to clothe God in new garb.

So, with hero worship badly tarnished, what's next? The way forward is to finish the work they began, which means showing exactly how the universe is behaving in a conscious way. It's a matter of providing evidence that everyone can agree upon, regardless of their built-in prejudices. Science exists to sort out the truth. Koalas and pandas both look like bears, for example. Yet both are vegetarians, which isn't bear-like, and neither species lives in areas where other bears also live. The matter couldn't be settled without incontrovertible evidence. The koala was straightened out first, because it carries its newborn young in a pouch and therefore isn't a bear but a marsupial like the kangaroo. The giant panda took a while longer, until genetics proved that it actually is a bear, and one of the most ancient species of bears. (Oddly, the giant panda has the genes of a carnivore instead of an herbivore, which means that it can extract very little energy from the bamboo leaves it feeds upon—so little, in fact, that the creature's activity is almost totally given over either to eating or sleeping. There's not even enough excess energy for males to fight over females in breeding season.)

So what kind of evidence would satisfy the everyday rational person (we'll exclude die-hard skeptics, who are beyond persuasion) that the universe is conscious? We will offer a sizable number of behaviors that do better than that. They indicate not just a conscious universe but a human universe. In such a universe, human beings find a true home, and at the same time our age-old dream of being completely free finally comes true.

SQUARE ONE IS NO PROBLEM

If there were a camp of biologists who thought that pandas are plants or koalas are insects, they wouldn't get past square one. In cosmology, there are basically two camps, "matter first" and "mind first," and they agree that square one is beyond spacetime, a dimensionless realm of nothing except pure potential. We've covered that pretty well already. Einstein pointed out that if the physical objects in the universe disappeared, there would be no space or time. As it blinks in and out of existence, every subatomic particle checks into the quantum vacuum, which means that it goes where time and space don't exist. The fact that the entire cosmos takes the same journey means that eternity is right beside us, a constant companion.

Another thing that both camps agree upon is existence, which sounds so basic as to be meaningless. Of course the universe exists. But that statement does have meaning, because it says that even when a particle takes its little trip into the quantum vacuum, the absence of time and space doesn't annihilate it. Somehow, the particle still exists, but it exists in eternity and everywhere at once. So powerful is the embrace of the quantum vacuum that when a quantum is behaving like a wave, it retains its ability to be everywhere at once. In short, existence isn't a blank slate. There's something valuable hidden in its secret recesses. (Some physicists, without a mystical blush, reduce the entire universe to a single wave or even a single particle. This would constitute the true God particle.)

Having agreed on square one, the next step is where argument enters the story. Was the infant cosmos pushed into existence by physical forces or by a mind? Is it enough to have bricks without a bricklayer? To illustrate, consider a cathedral in place of the universe. Studying the materials that the great Cathedral of Notre Dame is made of, such as stone, metals, and stained glass, can give hints about the building's construction methods and the historical times during which it was built; but by no means is Notre Dame merely the sum of these parts. It was created by conscious beings and reveals an alive presence that "dead" physical objects cannot account for. Stone, metal, and stained glass are the materials of architecture but not its art. So when it comes to describing Notre Dame, the parts tell us about the quantity of "stuff" a cathedral is made from; the architecture tells us about the qualia of the building, including its beauty and religious significance. Closing this gap between quantity and qualia would get us to step two of discovering the "real" reality of the universe.

We need a bricklayer who functions for science the way God

functions for religion. The universe has infinitely more complex building blocks than a cathedral, and the only candidate for a bricklayer who can keep them all straight is the cosmic mind. With Notre Dame, the presence of consciousness is unmistakable, even though the architects are long dead and gone. Inference is enough to tell us that conscious agents were at work. You can infer the behavior of consciousness in the cosmos the same way, through inference—there is no need to meet and greet a cosmic architect. We only need to observe how the universe is behaving, not like bits of matter colliding but like a mind doing everything with a purpose.

THE HUMAN TOUCH

If you declare that consciousness has no place in explaining how the universe works, the human mind is left hanging out on an evolutionary limb by itself. Is that really probable? Some physicalists will reluctantly concede mind-like behavior in the cosmos while refusing to call it conscious, finding the word radioactive. It is thought that soon after the big bang, much of creation was obliterated as matter and antimatter annihilated each other. But a tiny imbalance in favor of certain constants allowed the visible universe to exist, implying that matter and antimatter can reach a kind of peace treaty before both sides were totally wiped out. This reconciliation is known technically as complementarity, and two opposites that find a way to coexist are said to be complementary. For example, when two particles are entangled, as physics calls it, they display mirrored characteristics like spin and charge even when separated by billions of light-years. This makes them complementary. A change in one particle is instantaneously mirrored in the other. The implication is that complementarity is more fundamental than relativity, which takes the speed of light as an absolute limit. Instant communication isn't allowed. And yet nonlocality occurs. This means that entanglement is more fundamental than the four basic forces in nature, which are bound by rules that also take the speed of light as a limit.

It's fascinating to imagine how particles separated by billions of light-years could be "talking" to one another, and yet the same mystery exists much closer to home. In the brain it takes the coordinated effort of neurons scattered here and yonder to produce the three-dimensional image we call the physical world. This kind of coordination is also instantaneous, just as it is with elementary particles. The entire scheme works as a whole. On a movie set, the director calls for lighting, photography, sound, and action. Each one is a separate setup, and coordinating them takes time. But when you look out at the world, the mind doesn't say, "I got the lights going. Where's the sound? Will somebody cue the sound, please?" Instead, there is instantaneous coordination of all the elements needed to produce the movie of life.

What this implies is that complementarity isn't a property of particles or matter in general. It's a property of consciousness; actually it is one of the most fundamental ways that consciousness manifests the universe. Which supports the "mind first" camp quite strongly. But if we keep piling on evidence for a conscious universe, is that enough to justify a human universe? Are we really positioned in the wheelhouse of creation, or are we worker bees obeying the commands of cosmic consciousness? The question is rhetorical, because the only consciousness we know or can possibly know is human. Every law of nature became known through the human nervous system. We are the measure of creation, not by divine decree but because of complementarity, which fits every aspect of nature into a scheme perfectly suited to human existence.

All the other alternatives trap us inside mind-made boundaries. These boundaries carry built-in traps. For example:

- If we perceive human beings as accidental winners in the multiverse casino, then our existence depends upon random chance.
- If we perceive ourselves as products of physical forces, then we are no better than robots made of organic chemicals.
- If we tell ourselves that we evolved through survival of the fittest, then we are just the beastliest of beasts.
- If we see ourselves as a complex construct of information, then we are just a bunch of crunched numbers.

CAN REALITY SET US FREE?

At its core, the story of humankind has been a story of expanding consciousness. That's been the case for millennia, and the story is far from ended. But at last we can answer the nine cosmic mysteries this book began with.

Mystery 1: What came before the big bang?

Answer: A pre-created state of consciousness, which has no dimension. In this state, consciousness is pure potential. Every possibility exists in seed form. These seeds are made of nothing that can be empirically measured. Therefore, to claim that there was nothing before the big bang is just as correct as saying that everything existed before the big bang.

Mystery 2: Why does the universe fit together so perfectly?

Answer: It doesn't, because "fitting together" would mean that separate parts have to be carefully jiggled into place. In fact, the universe is one undivided whole. Its parts, whether we are talking about atoms, galaxies, or forces like gravity, are just qualia—the qualities of consciousness. All qualia exist on the same playing field as far as reality is concerned. You go to the same place to see the image of a rose in your mind's eye that nature goes to when it creates an actual rose.

Mystery 3: Where did time come from?

Answer: The same place that everything comes from, consciousness. Time is a qualia, like the sweetness of sugar or the colors in a rainbow. All are expressions of consciousness once the universe was hatched from the womb of creation.

Mystery 4: What is the universe made of?

Answer: The real building blocks of the universe are qualia. There is room for infinite creativity depending on the observer. The state of awareness that you are in alters the qualia all around you. A sunset isn't beautiful to someone who feels suicidal; a severe leg cramp is negligible if you've just won a marathon. Observer, observed, and process of observation are intimately linked. As they unfold, the "stuff" of the universe emerges.

Mystery 5: Is there design in the universe?

Answer: The answer is trickier than yes or no. If there was design "in" the universe, the two would have to relate the way a potter and a lump of clay relate. Form would emerge from the formless by applying an external mind. A familiar homily in Christianity refers to the human body in this way, as the vessel of God. In reality, design is a conscious perception that is totally malleable. One person can regard a wildflower as a thing of beautiful design while a second person sees it as a weed or a neutral biological specimen. After they vacate the meadow, a gopher may perceive the wildflower as food. Design is the interaction between mind and perception. It is permissible to see the universe as perfectly designed, perfectly random, a mixture of the two, or, as some mystics would declare, mere dream stuff with no substantiality at all.

Mystery 6: Is the quantum world linked to everyday life?

Answer: This one is also a bit tricky. The qualia of experience change depending on your state of awareness. In our normal waking state, the quantum domain is too small to be experienced directly, and linking it to the world of large objects proves very difficult. With no experience to guide us and with conflicting conclusions from laboratory experiments, physical linkages are controversial. But if you accept that the quantum domain isn't just mind-like but represents mind taking on the appearance of quanta, then the answer is relatively simple. The quantum domain is another qualia realm like any other. It needs no link to everyday life because all domains are constructed from consciousness. But a direct experience of the quantum realm is prevented by veiled nonlocality and cosmic censorship.

Mystery 7: Do we live in a conscious universe?

Answer: Yes. But this won't make any sense if your notion of a conscious universe is filled with thoughts, sensations, images, and feelings. Those are the contents of the mind. Remove the contents and what remains is pure consciousness, which is silent, unmoving, beyond time and space, yet filled with creative potential. Pure consciousness gives rise to everything, including the human mind. In that sense, we don't live in a conscious universe the way renters occupy a rental property. We participate in the same consciousness that *is* the universe.

Mystery 8: How did life first begin?

Answer: As a potential in consciousness that grew from seed form into every variety of living thing. Choosing to call the soft green moss on a rock a living thing while denying life to the rock is merely a mind-made distinction. In reality, everything in existence follows the same path from its origin (dimensionless being) to a state that consciousness chooses to create out of itself. Since they follow the same path from the unmanifest to the manifest, a rock and the moss that clings to it share life on equal terms.

Mystery 9: Does the brain create the mind?

Answer: No, but the reverse isn't true, either—the mind doesn't create the brain. This is another example of putting a distance between a potter and a lump of clay. Mind and brain aren't related in that way. Mind didn't find some primal stuff lying around in intergalactic space and fashion it into a brain. Matter didn't gather into bigger, more complex clumps until they got complex enough to begin thinking. The principle that applies here is complementarity, by which apparent opposites cannot exist without each other. There is no chicken-or-the-egg dilemma, because reality creates opposites all at once.

Being realistic, these answers sound very dissimilar to the answers you probably expected. We are quick to add, however, that nothing we've said is anti-science. What brought science to the end of its empirical methods wasn't a conspiracy of mystics, poets, dreamers, sages, and odd misfits. The everyday methods of science were outmoded by reality itself. In a universe dominated by dark matter and energy, where time and space break down at the Planck scale, it's not anti-scientific to look for a new way forward.

We've put three cards on the table: qualia, consciousness, and the human universe. What game will be played with them? No one can predict. The most brilliant insights into consciousness that inspired the quantum pioneers have lain fallow for almost a century. Taking the physical universe at face value remains the default mode, with a few exceptions.

In the end, we've been telling you about a hidden reality. It wasn't hidden intentionally or for mischievous purposes. The mind forged its own manacles, and it would take the history of the world to explain why and how.

Happily, the urge to know reality can never be eradicated, and something inside us, whoever we happen to be, yearns to be free. It was a fateful day when Einstein sat down with a mystical Indian poet to wrangle out the true nature of existence. If Tagore was right that the human universe is the only one that exists, we face a future of infinite hope in the joy of creation. For future generations, "You are the universe" will be a credo to live by, no longer a dream wrapped inside a mystery.

APPENDIX 1

GETTING COMFORTABLE WITH QUALIA

For many readers the term *qualia* will be new and perhaps alien. We have placed so much importance on this word that we want you to be comfortable with it. One difficulty is that qualia are all-inclusive: every experience is made of qualia, or qualities in consciousness. On a nice summer day it's not hard to accept the qualia delivered by the five senses—the warm air, bright sunlight, the smell of newly mown grass, and so on.

It's harder to believe that your body is also experienced as qualia. All the sensations you are having at this very minute would have no reality unless you experienced them personally, and therefore the body is a bundle of qualia. Going a level deeper, the brain's experiences are also qualia. When a concept becomes this universal, it's hard to know what to do with it. Where are the rules and boundaries, or do we live in a reality made of qualia soup? And what about the experience of an external reality, a "world out there"? That is also a qualia experience.

There are no rules to qualia that have the same status as the natural laws that classical physics laid down and that quantum physics took to an unimaginable level of sophistication. A ripe, sweet peach floods the senses with experience, not numbers, equations, and principles. One can't use the same vocabulary as in the domain of physicalists. "Sweet" isn't heavier, lighter, bigger, smaller, or denser than "ripe" or "warm."

The great advantage of qualia science, if that's the direction science takes in the future, is how perfectly it matches reality. Tasting a peach is a direct experience, needing no conceptual framework. This very absence of abstract concepts greatly irritates many mainstream scientists, but it's the seed for a new view of nature, transforming the physical universe into a consciousness-based universe.

To give you a compact vision of how qualia science might develop in the future, here's a brief set of principles, which we have distilled from the expanded argument of this book.

QUALIA PRINCIPLES

The Foundation for a Science of Consciousness

1. Science is materialistic, accepting as a given that the physical universe exists as it presents itself. But quantum physics long ago undermined the very notion of physical objects—at its foundation, the universe isn't solid, tangible, or fixed. Therefore, the old science of an external physical universe has been mortally wounded by the new science of quantum physics.

2. This ambiguity opens the door for an entirely new interpretation of nature: qualia science.

3. If physicality is radically compromised, what can be taken as a reliable foundation for future science? The one constant that is rejected by materialists: consciousness. Consciousness makes all experience possible. Attempts to exclude it from "objective" experiments cannot elude this fact.

4. Qualia science begins with the assertion that consciousness is not a trait that evolved from a material basis until it fully

emerged in human beings. Consciousness is fundamental and without cause. It is the ground state of existence. As conscious beings, humans cannot experience, measure, or conceive of a reality devoid of consciousness.

5. Consciousness, as the ground state of "normal" reality, behaves like a field, in all respects like quantum fields for matter and energy. As in any field, consciousness interacts with itself. This interaction proliferates into every specific form of consciousness, such as our own. (Consciousness did not arise as a secondary property of atoms and molecules over time.) But it must be understood that there is a deeper level of consciousness that has no dimensions, because any dimension in spacetime contains qualia, and in itself, pure consciousness has no qualia—it is the source of qualia, just as the quantum vacuum is the source of quanta. Consciousness can be considered the field of all fields, since it is the field that makes possible the existence of all fields.

6. Every specific form of consciousness (an elephant, porpoise, rhesus monkey, or a person) experiences the world subjectively. Individual subjectivity remains within the field of consciousness, which is its source. No form of consciousness is isolated from its source, just as no electromagnetic activity is ever isolated from the universal field of electromagnetism.

7. For human beings, subjective experiences are in the form of sensations, images, feelings, and thoughts (SIFT). The general term for these is *qualia*. Subjective reality is a vast composite of different qualia, such as color, light, pain, pleasure, texture, taste, memory, desire, anxiety, and joy.

8. All subjective experiences are qualia. This includes every perception, cognition, and mental event. No mental event can be

left out, including feelings of love, compassion, suffering, hostility, sexual pleasure, and religious ecstasy. At a subtle level, qualia are perceived as insight, intuition, imagination, inspiration, creativity.

9. "Objective," external physical reality comes to us through the qualia we are set up to perceive, not in and of itself. Without our subjective participation, space, time, matter, and energy, including all scientific variables and quantities, do not exist in themselves—or if they do, their reality is impenetrable. We live in a qualia universe. All our interactions with it are experiential, hence, ultimately, subjective. (Objective data have no independent existence, since they must be part of the data collector's experience.)

10. The experience of the body is a qualia experience. The experience of mental activity is a qualia experience. The experience of the world—and any other worlds—is a qualia experience.

11. The "I" feeling is a qualia experience. The "you" experience is a qualia experience.

12. Qualia, then, allow us to connect everything together through a common property—every single thing is an aspect of one field of consciousness.

13. As conscious beings processing reality every moment of our lives, we express ourselves in a qualia vocabulary. The qualia vocabulary is an attempt to put experience into words. The language of science, however, attempts the opposite: to extract experience in the name of objectivity. But "objectivity" itself denotes an experience. A language set apart from qualia doesn't exist.

14. Other life-forms, such as insects, bacteria, animals, and birds, have their own qualia niche. It is inaccessible to us—although we can try to imagine it—because each species has its own nervous system; even microorganisms have responses to the environment (seeking light, air, food, and each other). Insofar as we can interpret any other life-form, we are only reflecting the human nervous system's qualia processing. In actuality, the reality perceived through other nervous systems cannot be known to us.

15. Perception is the engine that creates species-specific experiences. Each experience reshapes physical reality, leading (in humans) to a qualia vocabulary that keeps up with every new change. The fact that "lower" animals, including insects and birds, also have extremely complex vocabularies is evidence for the creative link between language and reality.

16. We do not see because we have eyes. We do not hear because we have ears. The organs of perception do not create perception but are the lens through which consciousness and its qualia create perceptual experience. The perceptual can never be the actual. We perceive what our species has evolved to perceive. Whatever is truly real—the actual—is more primal than the things we perceive or think about or sense. Qualia science explores the boundary between the perceptual and the actual, with the goal of crossing over it.

17. The human brain represents the reality perceived by a particular life-form. Experience isn't organized randomly but symbolically. We humanize reality and in turn the qualia that register in the brain (pain, light, hunger, emotions, etc.) cause the brain and body to evolve as symbolic representations. This feedback loop originates in consciousness, not in the biology of the brain. Human consciousness is a specific expressive outlet for

the undifferentiated field of consciousness—the one creates the many.

18. Although we can interact with other life-forms like dogs and birds, we cannot presume that their qualia experience is identical to ours. What feels hot, cold, light, heavy, slow, fast, and so on to another species is unknowable—we cannot presume to say that these basic qualia register for them in a way similar to our response. We infer that they have feelings and sensory experience similar to ours, but that's the most we can say. It's very unlikely that the caw of a crow sounds the same to a crow as it does to us, or the bark of a dog to a dog. Yet we can communicate with each other as humans because we translate our qualia signals into a qualia vocabulary that is generally accepted (despite wide variances from person to person and culture to culture).

19. Each living entity creates its own perceptual reality by interacting with the fundamental ground of existence, pure consciousness. Pure consciousness is a field of all possibilities. Each possibility emerges, when it does, as qualia. The field of pure consciousness, however, exists prior to qualia; it is indescribable and inconceivable by a brain that knows reality only through qualia. The womb of creation is beyond space, time, matter, and energy.

20. There are as many perceptual realities (physical brains, bodies, worlds) as there are living entities with their qualia repertoire.

21. Our understanding of subjective experience, or our sense of empathy with others, takes place through the resonance of shared qualia. Whatever insight and connection we have to other species, beings, or realms of existence takes place through the

sensitivity and refinement of our subjective qualia in relation to their subjective qualia. What we call empathy is a shared resonance that registers in awareness.

22. Birth is the beginning of a particular qualia program. An individual qualia entity emerges into the world with a potential in qualia that unfolds as life. What happens over a lifetime is what we have in common: namely, interacting with other qualia entities and their qualia programs.

23. Death is the termination of a particular qualia program (the life program of an individual). The qualia return to a state of potential forms within consciousness, where they reshuffle and recycle as new living entities.

24. The consciousness field and its matrix of qualia are nonlocal and immortal. Nonlocal means that the field is all-pervasive and everywhere the same. (In fact, the very term *everywhere* is itself a qualia.) A field is affected by every specific event that happens in it. The whole never loses contact with its parts; they are never lost or forgotten.

25. We do not experience the field itself but the qualia that emerge from it. We use these to become individuals with a specific (i.e., local) perspective. Locality is a qualia experience in the nonlocal consciousness field.

26. Quantum mechanics is a mathematical model for measuring qualia mechanics, defined as our set of experiences of nature. It's the map, not the territory. At bottom, the map is mathematical because the quantum domain exhibits precise forms and probabilities. Mathematics leads to data, reducing experience to numbers. As such, this way of mapping reality loses all the qualia that constitute experience.

27. Reality can be mapped to resemble what it actually is—a continuous, dynamic flow of consciousness, emerging from the universal field and differentiating into matter, energy, worlds, and beings. Capturing what really exists, as opposed to the numbers that measure it in small, frozen slices, demands that science be revamped into qualia physics, qualia biology, qualia medicine, and so on.

28. The ancient wisdom traditions in many cultures recognized that subjective knowledge is useful and organized. These traditions take the qualia world and organize it into principles and behaviors of consciousness. Consciousness has recognized reference points, which is how Ayurveda, Qi Gong, and other qualia-based medicine became orderly, reliable, and efficacious. Even in Western materialism, room has been made for psychology, schools of psychotherapy, mythology and archetypes, childhood development, and gender studies—all of these branch out from subjective (qualia) experience of the world.

29. Spiritual practices are not unique or set apart from everyday experiences. They are based on subtle reference points in consciousness—in effect, they map self-awareness. Human consciousness looking at itself is a mirror for the field of consciousness looking at itself.

30. Spiritual practices fine-tune self-awareness. When the tuning is fine enough, qualia no longer mask where they come from. This is like seeing the mirror instead of the reflection. Consciousness sees itself and recognizes its pure, absolute existence—the pre-created state. Even when the world's wisdom traditions have degraded, losing a solid connection to pure consciousness, there are instructive relics of the old qualia science, which being alien to modern science, gets interpreted as the paranormal, miracles, and wonders. In fact, the supernatural

doesn't exist except as a subtler aspect of nature unfolding in qualia. These outside-the-normal qualia have as much legitimacy as the qualia that science has stamped with respectability.

31. Qualia medicine has already emerged in diverse forms around the world, such as Ayurveda and Traditional Chinese Medicine (TCM). Besides providing a fund of knowledge about the working action of herbs, these ancient traditions demand modern research to determine scientifically how the body responds, not just to herbs, but to every influence in the environment. The field of epigenetics has begun to flourish, examining how everyday experiences and stresses alter genetic activity.

32. Qualia biology would lead to a new understanding of life and its origins. Life has always existed as pure consciousness. Every property that has emerged in living things had its source as unmanifest potential, primary intelligence, creativity, and the evolutionary impulse. Being nonlocal, the field of infinite possibilities has no beginning. Therefore, life has no beginning either. What begins, evolves, declines, and ends are life-forms carrying out their qualia programs.

33. The origin of life-forms is the differentiation of pure consciousness (pure life) into multiple forms of life, or qualia conglomerates (life in the relative world).

34. The evolution of species is through natural selection, but in a much more comprehensive sense than Darwinian natural selection, which is based entirely on breeding rights and finding food for survival. What members of a species actually select for is enhanced qualia experience; this is the driving force in evolution, and since consciousness is unlimited, new qualia emerge, flourish, and seek maximum expression. The wild variety of life on Earth is a collective attempt to turn one planet's ecology into

a playground for qualia. The purpose of evolution is to maximize experience of every kind.

35. Evolution is purpose-driven through each species as it experiments with its environment and gets feedback. A feedback loop is set up that creatively meets every challenge from the environment, sometimes successfully, sometimes not. When seen as a whole, life on Earth is a qualia network, but so are the individuals within each species—everyone's experience affects the whole.

36. Genes, epigenes, and neural networks store and remember each step of evolution, following the path traced by experience. Seen for what they actually are, these recording devices are symbolic signatures of dynamic qualia networks. Each network is self-organizing, because no two species and no two individuals are working from exactly the same qualia program. Each scenario is unique; each works through its own possibilities.

37. Evolution is a never-ending process because it is rooted in an inherent property of consciousness, the impulse to create. Although evolution is synonymous with growth, the actual process includes the preservation of new creations and absorbing them into the entire system, whether that system is the human body, a niche in the environment, or the entire cosmos.

38. Humans have the gift of self-awareness, which is the key to freedom. Self-awareness means that we are not driven, much less are we imprisoned, by our qualia propensities. We are as dynamic as mind itself. This bespeaks an unbreakable connection with pure consciousness, which by definition cannot be a prisoner of itself. Infinite potential knows no limitations. Self-awareness, coming to terms with its true nature, will be the starting point for the next leap in our creative evolution as a spe-

cies. This leap will also remake the cosmos, since we inhabit a humanized universe. The universe fits our perception of reality.

39. This leap in evolution will be conscious, dictated by human aspirations. It will involve the emergence of new self-organizing networks of qualia structures and conglomerates. That is, a new mind-set will emerge, catch fire, reach a tipping point, and finally establish itself as the next human reality. Such a transformation isn't mystical. When the layers of aggression, war, poverty, tribalism, fear, deprivation, and violence begin to fall away, the qualia that remain lie closer to their creative source. It's crucial for outworn qualia to be peeled away first; in turn, this requires that inertia, born of unconsciousness, is abandoned in favor of the dynamic growth of new qualia networks.

40. Quantum mechanics and classical science will always be useful for the creation of new technologies, but qualia science could take our civilization in the direction of wholeness, healing, and enlightenment.

APPENDIX 2

HOW COSMIC CONSCIOUSNESS BEHAVES

Modern physics has given us a detailed picture of how the physical universe behaves. The only problem is that there is no purpose or meaning to the picture. If we want to topple the reliance on randomness as the prime mover in the cosmos, we need to take the same picture and show what, if anything, is added by introducing the cosmic mind.

Here, in brief, are the actions of consciousness in the universe, each one chosen to address known behaviors throughout creation, behaviors founded on quantum principles.

1. Cosmic consciousness keeps opposites in balance without one side abolishing the other. The coexistence of opposites is called *complementarity.* In any situation in which opposites exist, one can replace the other in specific circumstances, yet at the same time each implies the other, as negative implies positive and north implies south.

2. Cosmic consciousness devises new forms and functions out of itself. This kind of self-organization is called *creative interactivity.* In living organisms, there is sentient interactivity: Living creatures constantly interact with their environment, including other sentient beings, seeking food, propagating the species, and

being aware of the existence of "others" at different levels. The argument that only human beings possess sentience is hollow—it's a basic attribute of consciousness itself.

3. Cosmic consciousness has the urge to build upon the old to create the new. This behavior is called *evolution*. Confining evolution to life on Earth is a narrow perspective. The entire cosmos exhibits evolution as a basic trait. The alternative—a universe operating randomly for more than 10 billion years, only to hit upon evolution when planet Earth appeared—is unreasonable. What brought the planets into existence if not evolution from simpler collections of matter?

4. Cosmic consciousness operates locally through separate events that are too far apart to be considered in touch with each other, but at the same time it holds these events together at a deeper level where nothing is separate. This trait is called *veiled nonlocality*.

5. Cosmic consciousness sets up the universe so that our way of viewing, whether through physics or biology, isn't violated. Each perspective justifies itself. No matter how many stories we tell about reality, the whole story is kept from view. This trait is called *cosmic censorship*.

6. All the parts of the cosmos are structurally similar or can be seen as having likenesses at deeper levels. Two observers looking at different levels of nature can communicate and understand each other because of repeated patterns and forms that share resemblances. This principle is known as *recursion*.

Cosmic consciousness mirrors the observer's state of being. There is no privileged point of view, even though in the past religion claimed to have a privileged point of view while today science does the same. But each story is provided with evidence to support it, because our state of being interacts so intimately with reality that observer, observed, and the process of observation are inseparable. What we've just outlined are the behaviors of every aspect of nature; they aren't metaphysical dreams. Cosmic consciousness produced the universe as a living, self-organizing system. At every instant since the big bang, nature keeps repeating the same behaviors at every level. In biology it is undeniable that living things organize themselves, using DNA as a basic template. Horses create baby horses; horse livers create new liver cells; each cell sustains the process of eating, breathing, excreting, dividing, and so on. This self-organization is dynamic, and when necessary, it has the flexibility to adapt to new conditions. A horse can live high in the Andes Mountains or below sea level in Death Valley because its cells are adaptable. A horse can run or stand still. It can be pregnant or not. These are massive changes of state, but the horse's body, from the level of its DNA upward, regulates itself. If it didn't adapt to changing conditions, it would die.

This ability to adapt is reflected in how a molecule is organized and an atom and a quark. In all cases there is adaptation in the face of change, and *the whole system participates.* If we scrutinize a horse at various levels, we see atoms, molecules, cells, tissues, organs, and finally the complete creature. But the horse is *more* than a collection of its parts, as a cathedral is more than glass, stone, marble, metal, cloth, and precious stones. If a horse's liver cells opt out, there can be no horse. If the DNA inside a cell decides not to divide, there is no horse. Why don't all kinds of things opt out? There are trillions of participating parts in a living horse. Cars and trucks have numerous parts, and much to our frustration, a few always seem to be breaking down or threatening to.

But so far as nature is concerned, a horse is only one thing, a species of awareness, and at the level of awareness, all participation is unified. For any living creature—a blowfish, fruit fly, or horseshoe crab—there is interconnection at each level. Each level retains its own integrity while meshing into the next level. This dynamic stream of cooperation is the modern equivalent of the religious notion of the Great Chain of Being, which held that God seamlessly wove together every level of creation. In nonreligious terms, we say that complex systems organize themselves through the natural behavior of consciousness, the behaviors we've just listed.

The following is a grand summation of the things that put human beings foremost in the universe. To understand this, you don't have to look through the Hubble telescope. Much closer to home, a heart, liver, or lung cell behaves like the universe itself. The matchup is perfect.

HOW EVERY CELL MIRRORS THE COSMOS

Complementarity: Each cell preserves its individual life while maintaining a balance with the whole body. Even cells that seem like opposites, such as a bone cell and a blood cell, are necessary to each other. They are necessary to the whole.

Creative interactivity: Each cell produces chemical products to fit specific situations, such as how much oxygen is needed in the blood at very high versus very low altitudes. Genes adapt creatively to change all the time by creating new mixes of chemical products in the cell.

Evolution: All cells begin with the same DNA as well as the same general stem-cell structure. In the womb these stem cells re-create the entire evolution of life on Earth, going through specific stages until the final evolutionary stage, becoming human, is reached.

Veiled nonlocality: Each cell has perfect knowledge about the events it controls, but the wholeness of the body is invisible and concealed. It has no physical fingerprint, even though the wholeness of the body is the whole point of every event taking place in a cell.

Cosmic censorship: Every cell mirrors the laws of biology, which cannot be violated—otherwise the cell would be unable to exist. What "censors" nonlocality or wholeness is the appearance of almost infinite events taking place all around us, seemingly following established reality but in fact veiling or clouding what lies "underneath" ordinary perception. In duality, even the mind can't know its own wholeness through thinking.

Recursion: As different as cells look when gathered into kidney, bone, heart, or brain tissues, they are basically the same. They follow the same patterns. (At the deepest layers of physicality, all electrons are the same, prompting Richard Feynman to state that there is really only one electron.) Recursion allows understanding to be built up from familiar patterns. We can understand one another and communicate. This is made possible by repeating the same processes in each cell and linking all of them back to DNA.

INDEX